Humidity Sensors - Types and Applications

Edited by Muhammad Tariq Saeed Chani,
Abdullah Mohammed Asiri
and Sher Bahadar Khan

Published in London, United Kingdom

IntechOpen

Supporting open minds since 2005

Humidity Sensors - Types and Applications
http://dx.doi.org/10.5772/intechopen.83288
Edited by Muhammad Tariq Saeed Chani, Abdullah Mohammed Asiri and Sher Bahadar Khan

Contributors
Avik Sett, Tarun K. Bhattacharyya, Kunal Biswas, Santanab Majumder, Arkaprava Datta, Jude Iloabuchi
Obianyo, Rajesh Kumar, Amine BelHadj Mohamed, Jamel Orfi, Ahmad Alfaifi, Adnan Zaman, Abdulrahman
Alsolami, Meliha Oktav Bulut, Ayşen Cire, Muhammad Tariq Saeed Chani, Sher Bahadar Khan, Abdullah
M. Mohammed Asiri

Notice
Statements and opinions expressed in the chapters are these of the individual contributors and not
necessarily those of the editors or publisher. No responsibility is accepted for the accuracy of
information contained in the published chapters. The publisher assumes no responsibility for any
damage or injury to persons or property arising out of the use of any materials, instructions, methods
or ideas contained in the book.

First published in London, United Kingdom, 2023 by IntechOpen
IntechOpen is the global imprint of INTECHOPEN LIMITED, registered in England and Wales,
registration number: 11086078, 5 Princes Gate Court, London, SW7 2QJ, United Kingdom
Printed in Croatia

British Library Cataloguing-in-Publication Data
A catalogue record for this book is available from the British Library

Additional hard and PDF copies can be obtained from orders@intechopen.com

Humidity Sensors - Types and Applications
Edited by Muhammad Tariq Saeed Chani, Abdullah Mohammed Asiri and Sher Bahadar Khan
p. cm.
Print ISBN 978-1-83968-565-1
Online ISBN 978-1-83968-566-8
eBook (PDF) ISBN 978-1-83968-567-5

We are IntechOpen,
the world's leading publisher of
Open Access books
Built by scientists, for scientists

6,200+
Open access books available

168,000+
International authors and editors

185M+
Downloads

Our authors are among the

156
Countries delivered to

Top 1%
most cited scientists

12.2%
Contributors from top 500 universities

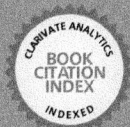

CLARIVATE ANALYTICS
BOOK
CITATION
INDEX
INDEXED

WEB OF SCIENCE™

Selection of our books indexed in the Book Citation Index (BKCI)
in Web of Science Core Collection™

Interested in publishing with us?
Contact book.department@intechopen.com

Numbers displayed above are based on latest data collected.
For more information visit www.intechopen.com

Meet the editors

Engr Dr. Muhammad Tariq Saeed Chani is an assistant professor at the Center of Excellence for Advanced Materials Research, King Abdulaziz University, Jeddah, Saudi Arabia. He received his Ph.D. in materials engineering from Ghulam Ishaq Khan Institute of Engineering Sciences and Technology, Pakistan, in 2012. His research interests are the fabrication and characterization of organic-inorganic semiconducting devices, which includes sensors, thermoelectric cells and solar cells. He has registered nine US patents. He has also published 75 research articles in various international journals.

Prof. Dr. Sher Bahadar Khan is a professor in the Chemistry Department, at King Abdulaziz University, Jeddah, Saudi Arabia. He obtained his Ph.D. from HEJ, Karachi University, Pakistan and worked as a post-doctoral fellow at Yonsei University, South Korea. His research in nanochemistry and nanotechnology concerns catalysis, hydrogen production, drug delivery, solar cells, the development of active photo-catalysts, and the fabrication of perceptive biosensors and chemi-sensors using metal oxide nanomaterials and their green environmental nanohybrids. He is the author of 350 research articles, 12 books, and seven patents with almost 2,000 ± 10 impact factor, 13,000 citations and a 65 h-index.

Prof. Dr. Abdullah Mohammed Ahmed Asiri is a professor and chairman of the Chemistry Department, at King Abdulaziz University. He is also the director of the Center of Excellence for Advanced Materials Research. He obtained his Ph.D. from the University College of Wales, Cardiff, UK, in 1995. He is the director of the Education Affairs Unit, Deanship of Community Services, and a member of the Advisory Committee for Advancing Materials, National Technology Plan, King Abdul Aziz City of Science and Technology. He is an editorial board member of the *Journal of the Saudi Chemical Society, Journal of King Abdul Aziz University, Pigment and Resin Technology Journal, Organic Chemistry Insights,* and *Recent Patents on Materials Science,* and is a member of several national and international societies and professional bodies.

Contents

Preface

The seven chapters of this book present various types of humidity sensors, sensing mechanisms, sensing materials, sensor design, fabrication technologies, sensor applications and other related studies. The first chapter provides a brief introduction to the topic. Chapter 2 describes the basics of humidity, the various types of humidity sensors, and humidity sensing techniques. Thick and thin film preparation processes are quite flexible and have advantages over other technologies. The third chapter reviews the different types of humidity sensors and applications, with an emphasis on fiber-optic, nano-brick, capacitive, resistive, piezoresistive and magnetoelastic humidity sensors. Fiber-optic sensors have been found to be best for use in harsh weather conditions, while nano-brick sensors have excellent humidity-sensing qualities. The chapter also compares capacitive sensors which use impedance with equivalent resistive sensors fabricated with ceramic or organic polymer materials and discusses the sensitivities of piezoresistive and magnetoelastic sensors.

Chapter 4 reviews MEMS humidity sensors made with microfabrication technologies and their operating principles. Capacitive humidity sensors are discussed, highlighting different sensing materials and the effect on the performance of their permittivity and physical parameters. Finally, piezoelectric and resistive humidity sensors, and their sensing mechanisms, are reviewed.

Chapter 5 looks at graphene-based humidity sensing devices, considering graphene synthesis methods, their mechanical and electronic properties, and sensing mechanisms and behavior. Recent trends in graphene, graphene oxide, graphene quantum dots, reduced graphene oxide, and graphene-composite-based humidity sensors are also discussed, together with the challenges and future trends of graphene-based humidity sensors.

Chapter 6 describes research on the water vapor permeability of polyamide 6.6/polyurethane fabric used for sportswear in the finishing process. The final chapter explores important aspects of falling film evaporation in several geometrical configurations such as on horizontal tubes and inside inclined or vertical tubes or channels.

We hope that the topics covered in this book will be relevant to students, researchers and general readers alike.

Muhammad Tariq Saeed Chani, Abdullah Mohammed Asiri and Sher Bahadar Khan
Center of Excellence for Advanced Materials Research,
King Abdulaziz University,
Jeddah, Saudi Arabia

Chapter 1

Introductory Chapter: Humidity Sensors

Muhammad Tariq Saeed Chani, Sher Bahadar Khan
and Abdullah Mohammed Asiri

1. Introduction

1.1 Humidity sensors: A brief introduction of materials, mechanism and classifications

Humidity refers to the presence of water vapor in the air. Humidity affects human health and physical qualities of materials [1, 2] and therefore, it is critical to measure and control the humidity. Humidity measurement and control are critical not only for human and animal comfort but also for manufacturing processes and industrial products [3–6]. Humidity measurement is considered very imperative in different industries such as health care, environmental monitoring, automobile, building air-conditioning, civil engineering, agriculture, semiconductor, pharmaceutical, textile, medical, paper, and process industries [7–9].

Humidity can be measured in three ways namely absolute humidity, specific humidity, and relative humidity. Relative humidity (RH) is one of the most commonly measured quantities in industry and everyday life [10]. RH is traditionally measured with microporous thin sheets and thin plates piezoelectric quartz sensor. The sensing mechanism of these materials is based on variation in the luminescence and oscillation frequency, correspondingly. These sensors may be expensive or demand high operational power/temperature, as well as a significant maintenance cost, depending on the nature of the materials. Humidity can also be measured using materials that indicate a change in resistance, impedance, or capacitance as a function of humidity. Such materials may be ceramics, low-molecular-weight organic materials, polymers, and composites. Humidity sensors are characterized as capacitive, oscillating, resistive, gravimetric, impedimetric, thermo elemental, hydrometric, or integrated optical based on their sensing method. The design of humidity sensors and nature of the material (sensing material) have an impact on their performance [10–14].

For a smart sensor, the required properties are the linear response, high sensitivity, wide sensing range, low hysteresis, fast response, high stability (physical and chemical), and low cost [14, 15]. To obtain these required features, several materials (organic, inorganic, and composites) have been studied. During last few decades organic-inorganic nanocomposites have been developed for advanced optic, magnetic and electronic applications. Despite having low stability as compared to inorganic materials, the organic materials have a lot of potential owing to their lightweight, high flexibility, high surface area, and easy fabrication. The merits of both materials (organic-inorganic) may be combined in a single device by using organic-inorganic composites as an active material [5, 10, 16–18].

There are various types of humidity sensors based on sensing mechanisms, sensing materials, sensors design, fabrication technologies, and applications. Humidity sensors can be fabricated by various types of fabrication techniques such as thick and thin film preparation processes, which are quite flexible and advantageous over other technologies. Similarly, micro-fabrication technology is also one of the useful fabrication techniques. Based on sensing mechanism, humidity sensors can be divided into capacitive, resistive, piezoresistive, magnetoelastic, field effective transistors, bulk acoustic wave, and optical humidity sensors. It has been reported that optical fiber sensors are best for use in harsh weather conditions, while the nano-bricks sensors and capacitive sensors, which use impedance and resistive sensors fabricated with ceramic or organic polymer materials have also excellent humidity sensing qualities. Several types of sensing materials are used for sensing humidity like different ceramic materials, resistive polyelectrolytic materials, conductive polymeric materials, low molecular weight organic materials, and graphene-based materials such as graphene, graphene oxide, graphene quantum dots, reduced graphene oxide, and graphene-composites. On the other hand, humidity sensors are classified in different classes, e.g. rigid, flexible, dynamic, and static humidity sensors. The efficiency of these humidity sensors can be increased by investigating and developing new sensitive, durable, and resistive environmental factors sensing elements for the humidity sensors. In this book, the main aspects of humidity sensors have been covered.

Author details

Muhammad Tariq Saeed Chani[1*], Sher Bahadar Khan[1]
and Abdullah Mohammed Asiri[1,2]

1 Center of Excellence for Advanced Materials Research, King Abdulaziz University, Jeddah, Saudi Arabia

2 Faculty of Science, Chemistry Department, King Abdulaziz University, Jeddah, Saudi Arabia

*Address all correspondence to: mtmohamad@kau.edu.sa

IntechOpen

References

[1] Camaioni N, Casalbore-Miceli G, Li Y, Yang MJ, Zanelli A. Water activated ionic conduction in cross-linked polyelectrolytes. Sensors and Actuators B: Chemical. 2008;**134**:230-233. DOI: 10.1016/j.snb.2008.04.035

[2] Chani MTS, Karimov KS, Khalid FA, Moiz SA. Polyaniline based impedance humidity sensors. Solid State Sciences. 2013;**18**:78-82. DOI: 10.1016/j.solid statesciences.2013.01.005

[3] Chani MTS, Karimov KS, Khalid F, Abbas S, Bhatty M. Orange dye— Polyaniline composite based impedance humidity sensors. Chinese Physics B. 2013;**22**:010701

[4] Chani MTS, Karimov KS, Khalid FA, Raza K, Farooq MU, Zafar Q. Humidity sensors based on aluminum phthalocyanine chloride thin films. Physica E: Low-dimensional Systems and Nanostructures. 2012;**45**:77-81

[5] Chani MTS, Karimov KS, Khan SB, Fatima N, Asiri AM. Impedimetric humidity and temperature sensing properties of chitosan-CuMn2O4 spinel nanocomposite. Ceramics International. 2019;**45**:10565-10571

[6] Fuke MV, Kanitkar P, Kulkarni M, Kale BB, Aiyer RC. Effect of particle size variation of Ag nanoparticles in polyaniline composite on humidity sensing. Talanta. 2010;**81**:320-326

[7] Farahani H, Wagiran R, Hamidon MN. Humidity sensors principle, mechanism, and fabrication technologies: A comprehensive review. Sensors. 2014;**14**:7881-7939

[8] Wang X-D, Wolfbeis OS, Meier RJ. Luminescent probes and sensors for temperature. Chemical Society Reviews. 2013;**42**:7834-7869. DOI: 10.1039/c3cs60102a

[9] Liu X, Jiang M, Sui Q, Geng X. Optical fibre Fabry–Perot relative humidity sensor based on HCPCF and chitosan film. Journal of Modern Optics. 2016;**63**:1668-1674. DOI: 10.1080/09500340.2016.1167974

[10] Chani MTS. Impedimetric sensing of temperature and humidity by using organic-inorganic nanocomposites composed of chitosan and a CuO-Fe3O4 nanopowder. Microchimica Acta. 2017;**184**:2349-2356

[11] Chani MTS, Karimov KS, Khan SB, Asiri AM. Fabrication and investigation of cellulose acetate-copper oxide nano-composite based humidity sensors. Sensors and Actuators A: Physical. 2016;**246**:58-65. DOI: 10.1016/j.sna.2016.05.016

[12] Chani MTS, Karimov KS, Bukhsh EM, Asiri AM. Fabrication and investigation of graphene-rubber nanocomposite based multifunctional flexible sensors. International Journal of Electrochemical Science. 2020;**15**:5076-5088

[13] Chani MTS, Karimov KS, Marwani HM, Rahman MM, Asiri AM. Electric properties of flexible rubber-based CNT/CNT-OD/Al cells fabricated by rubbing-in technology. Applied Physics A. 2021;**127**:236. DOI: 10.1007/s00339-021-04382-3

[14] Chani MTS, Karimov KS, Meng H, Akhmedov KM, Murtaza I, Asghar U, et al. Humidity sensor based on orange dye and graphene solid electrolyte cells. Russian Journal of Electrochemistry. 2019;**55**:1391-1396

[15] Asiri AM, Chani MTS, Khan SB. Method of making thin film humidity sensors. 2018: US Patents, US 9,976,975

[16] Chani MTS. Fabrication and characterization of chitosan-CeO2-CdO

nanocomposite based impedimetric humidity sensors. International Journal of Biological Macromolecules. 2022;**194**:377-383. DOI: 10.1016/j. ijbiomac.2021.11.079

[17] Chani MTS, Karimov KS, Asiri AM. Carbon nanotubes and graphene powder based multifunctional pressure, displacement and gradient of temperature sensors. Semiconductors. 2019;**53**:1622-1629

[18] Chani MTS, Karimov KS, Asiri AM. Impedimetric humidity and temperature sensing properties of the graphene–carbon nanotubes–silicone adhesive nanocomposite. Journal of Materials Science: Materials in Electronics. 2019;**30**:6419-6429

Chapter 2

Solid State Humidity Sensors

Rajesh Kumar

Abstract

A variety of humidity sensors have been developed to address the problem of humidity measurement in instrumentation, agriculture and systems which are automatic. Various types of humidity sensors have been reviewed along with their mechanisms of humidity detection. Thin and thick film preparation processes are quite flexible. This flexibility provides advantages over other technologies. After comparing all the aspects of different humidity sensors, it has been observed that there are still some shortcomings left, which need to be removed to enhance the humidity sensing capability, recovery and response times of the sensor elements.

Keywords: Humidity sensors, Relative humidity, thick/thin film, Fabrication technologies, capacitive/resistive sensors, protonic conduction mechanism

1. Introduction

Significant improvements have been seen in the sensor technology in recent years. The miniaturisation process provides a wide range of advantages to the field of sensor technology [1–11]. It is a well known fact that humidity plays a significant role in all the processes occurring on this planet. For the high efficacy of all these processes, the monitoring, detection and control of the humidity of the surroundings is of utmost importance [12, 13]. For the fabrication of good humidity sensors, choice of fabrication technologies, optimisation of the surface for conductance and cost are the major factors that play a very important role [14–27].

1.1 Basics of humidity

"Humidity is defined as the amount of water vapour in an atmosphere of air or other gases". The units of humidity parameters depend on the technique used. In this respect, "Relative humidity (RH)", "Parts per million (PPM)" by weight or by volume and "Dew/Frost point (D/F PT)," are used.

In addition to the above units, it is worthwhile to mention three more parameters and their relationships here.

    i. Absolute Humidity: It is defined as a ratio of the mass of water vapour in air to the volume of air. It is also called vapour density and its units are g/m^3 or $grains/ft^3$. The absolute humidity is given by

$$AB = m/V, \text{ where 'm' is the mass of water vapour and 'V' is the volume of air.} \quad (1)$$

    ii. Relative Humidity: It is defined as ratio of the amount of moisture content of air to the maximum (saturated) moisture level that the air can hold at a same given temperature and pressure of the gas. It is denoted by RH and depends upon the temperature and is a relative measurement.

RH (in percentage) $= Pv \times 100 / Ps$, where Pv is the actual partial pressure of the moisture content in air and Ps is the saturated pressure of moist air at same given temperature. $\qquad$ (2)

Saturation Humidity: It is defined as the ratio of the mass of water vapour at saturation to the volume of air. It is denoted by SH.

$$SH = m_{ws} / V \qquad (3)$$

SH is the saturation humidity (g/m^3), m_{ws} is the mass of water vapour at saturation (g) and V is the volume of air (m^3). The saturation humidity is a function of temperature. SH is a function of temperature and can provide the maximum amount of moisture content (mass) in a unit volume of gas at a given temperature. The percentage relative humidity can also be expressed as.

$$RH \left(In \ percentage\right) = AB \times 100 / SH \qquad (4)$$

Parts per million by volume (PPMv) is defined as volume of water vapour content per volume of dry gas and parts per million by weight (PPMw) is obtained by multiplying PPMv by the mole weight of water per mole weight of that gas or air. PPMv and PPMw are the absolute humidity measurements.

Dew point is defined as a temperature (above 0°C) at which the water vapour content of the gas begins to condense into liquid water, and Frost point is the temperature (below 0°C) at which the water vapour in a gas condenses into ice [28, 29]. D/F point parameters depend upon the pressure of the gas but are independent of the temperature. The ambient relative humidity is given by.

Ambient relative humidity $=$ Ambient temperature $-$ Dew point temperature $\quad$ (5)

2. Classification of humidity sensors

Depending upon the different operating conditions, a variety of humidity sensors have been developed in due course of time. Based on the units of measurement, absolute humidity and relative humidity, humidity sensors have been divided into two classes, which are explained as follows:

i. Relative humidity (RH) sensors

ii. Absolute humidity sensors (hygrometers)

It is worthwhile to mention here that relative humidity sensors are preferred over absolute humidity sensors. The RH sensors are classified into three classes: Ceramic type (based on semiconducting materials), organic polymer based sensors and organic/inorganic hybrid sensors (based on polymers and ceramic materials). All the above mentioned categories make use of changes in the physical and electrical properties of the sensor elements, when exposed to different atmospheric humidity conditions of the surrounding environment. They provide a measure of humidity depending upon the adsorption and desorption of water molecules.

In hygrometer type of sensors, humidity measurement is determined by either measuring the conductance or capacitance of the sensing material, when it is exposed to the environmental humidity [30–33].

The first electrolytic humidity sensor was developed by Dunmore based on Lithium Chloride (LiCl) in 1937. A porous supporting material was immersed in a humidity sensitive partially hydrolysed polyvinyl acetate which was impregnated with LiCl solution and a potential difference was applied across the supports to form an electrolytic cell. By absorbing the water vapours via the porous medium, the ionic conductivity of the cells was changed and humidity was detected [34–36]. These types of sensor elements suffered from the following types of drawbacks, which led to the development of impedance-sensitive humidity sensors [37–41]:

- Low response times

- Low recovery times

- Not reliable to work in conditions, which has high moisture content

The organic polymer film humidity sensors can be divided into resistive and capacitive types [42–44]. The resistive type sensors can be divided into electronic and ionic conduction type. In the electronic conduction type of sensors, polyelectrolytes respond to water vapour variations by changing their resistivity. On the other hand, in the ionic conduction type, variation of the dielectric constants of the polymer dielectrics changes the capacitance of the material and hence the humidity is measured.

Ceramic type humidity sensors based on metal oxides have clear advantages over their other counterparts such as good mechanical strength, thermal capability, physical stability and resistance to corrosive chemicals.

Depending upon the sensing mechanisms, ceramic type sensors can be divided into:

- Impedance type

- Capacitive type

The impedance type of sensors can be further subdivided into ionic conduction and electronic conduction types and work by observing the changes in the conductivity of sensor elements, when they are exposed to different levels of humidity. [45, 46].

The metal oxide ceramics elements are prepared by various conventional and advanced ceramic processing methods, with an aim to produce porous structures that support adsorption of water vapours from the surrounding atmosphere.

3. Principle of protonic-conduction type ceramic humidity sensors

A variety of humidity sensing mechanisms have been proposed by various research investigators for the humidity sensing by ceramic based sensors. The mechanisms of humidity sensing via ionic conduction, electronic conduction, solid electrolyte and capacitive types are based on water vapour adsorption either by chemisorption, physisorption or capillary condensation processes. As water molecules are absorbed on the surface of the sensor elements, so it is worthwhile to discuss about the hydrogen ion (H^+) and hydroxide (OH^-) ion diffusion into the sensor elements [47–51].

3.1 Hydrogen (H⁺) ions diffusion

This proton transfer mechanism was first proposed by Grotthuss. In this mechanism, protons are tunnelled from one water vapour molecule to the next water vapour molecule through hydrogen bonding as shown in the **Figure 1**.

In the proton carrier mechanism of ceramic humidity sensors, the adsorbed water molecules condense on the thin film surfaces or in bulk and protons carry out the conduction process [52, 53].

3.2 Diffusion and mobility of hydroxide (OH⁻) ions

On the similar lines of proton transfer, Grotthuss suggested another mechanism for the hydroxide ion transfer. According to this mechanism, the mobility of the hydroxide ions also occurs through proton transfer mechanism as shown in the **Figure 2**.

An auto ionisation reaction of the water vapours occurs on the surface of the sensor element due to amphoteric nature of water molecule. In this process, a water molecule becomes hydroxyl ion with the loss of a hydrogen ion. The released hydrogen ion donates the proton to another water molecule to form hydronium ion with has a formula H_3O^+.

At low humidity levels, the charge carriers are predominantly protons. The protons are transferred through hopping of hydrogen ions between the sites that have hydroxide ions present on them. At higher humidity levels, as more water layers are present, more dissociation of water molecules occurs to produce hydronium ions. When the thin film surface is completely covered with water molecules, diffusion of hydronium ions on hydroxide ions dominates. In addition to this, the proton transfer by hydroxide ions between the adjacent water molecules also occurs. This charge carrying continues when hydronium ion transfers a proton to a neighbouring water molecule and forms another hydronium ion. The mechanism involves the

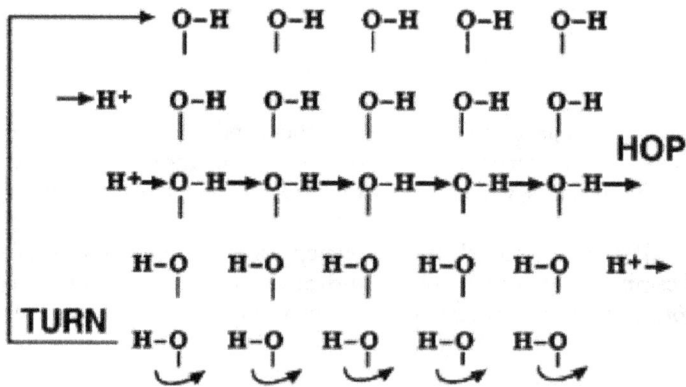

Figure 1.
Proton conduction of the hydrogen bonded networks between water molecules [54].

Figure 2.
Proton transfer mechanism of hydroxide ions.

dancing of protons from one water vapour molecule to the another water molecule. This leads to the change in resistance and capacitance of the thin film elements [55].

3.3 Water adsorption and conduction mechanism on ceramic oxide solid surfaces

It is well known that the surfaces of most of the metal oxides are covered with hydroxyl groups, when exposed to the humid atmospheres. This results in the formation of hydrogen bonds that facilitates the absorption of water molecules by ceramic oxide surfaces as shown in **Figure 3**. The formation of these hydrogen bonds results in the change of electrical conductivity of the surfaces.

Interaction of water molecules with ceramic oxide surfaces is a three step process, which are explained below:

In the first step, a few water vapour molecules are chemically adsorbed at the neck of crystalline grains on the activated sites of the surface. This leads to the dissociation of water vapour molecules to form hydroxyl groups. In this interaction, protons are generated which migrate from one site to another site on the surface and react with oxygen ions to form a second hydroxyl group as shown in the **Figure 4**.

In the second stage, more water vapour layers are physically adsorbed on the first stage formed hydroxyl layer, forming multilayers. The multilayers are more disordered than the first monolayer.

In the third stage, with the formation of more layers, a large amount of water molecules are physisorbed on the necks and flat surfaces. Therefore, singly bonded water vapour molecules have higher mobility and form continuous dipoles and electrolyte layers between the electrodes. This results in the increased dielectric constant and bulk conductivity. The multilayer formation of water vapour molecules

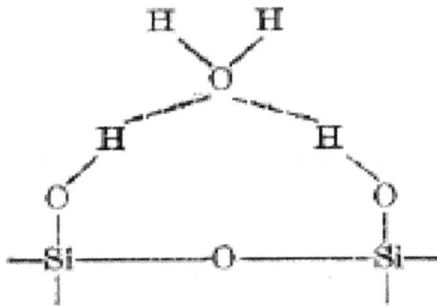

Figure 3.
The above diagram shows the adsorption sites on the silica surfaces and formation of hydroxyl pairs to hold water molecules.

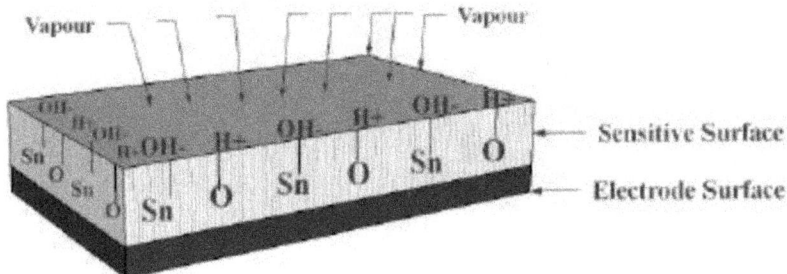

Figure 4.
Illustration of water vapour chemisorptions and hydroxyl layer formation on the surface of tin oxide.

on ceramic surfaces is confirmed by the increase of dielectric permittivity of the surface and is shown for iron oxide in the given **Figure 5.**

It has been found out that physisorption of water occurs at temperatures <100°C while chemisorptions occurs in the temperature range of 100-400°C. In this temperature range, hydroxyl groups interact with the surface of ceramic material [56–69].

The porous structure of the ceramics is mainly responsible for the physisorption of water vapours on the surfaces of humidity sensors, which are based on ceramic materials [70, 71]. For the cylindrical pores in the ceramic materials, the radius of the pores is given by the Kelvin Equation [72].

$$r_k = 2\gamma M / \rho RT \ln (p_s/p) \qquad (6)$$

r_k is called Kelvin radius of the cylindrical pores. γ,M and ρ are the surface tension, molecular weight of water and density respectively. p is called the vapour pressure of water and p_s is vapour pressure of water at saturation. R is called gas constant and T is called the absolute temperature [73].

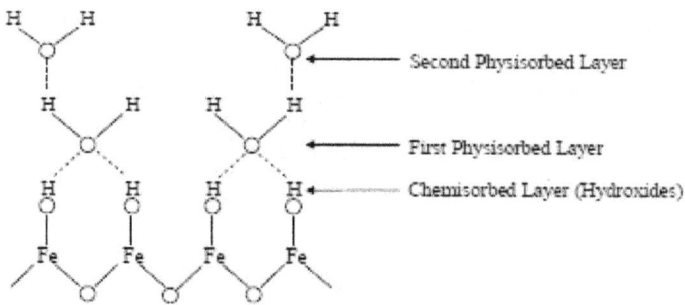

Figure 5.
The above diagram shows the multilayer structure of adsorbed water vapour molecules on the surface of iron oxide.

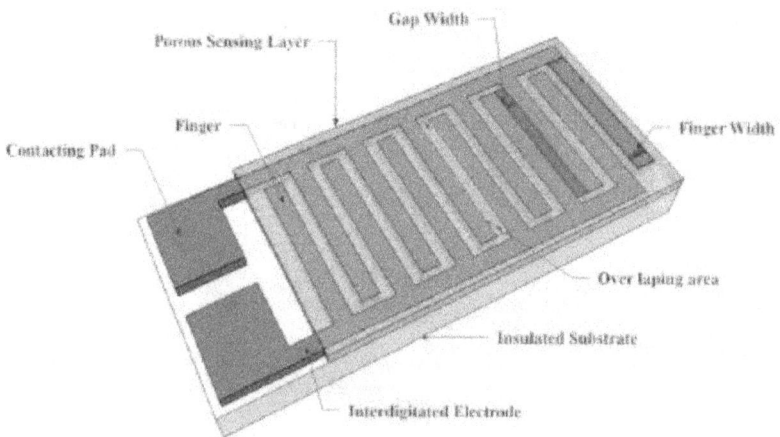

Figure 6.
The above diagram shows the planar thick/thin film based humidity sensor based on interdigital structure with porous sensing element.

3.4 Impedance type humidity sensors

These type of humidity sensors contain noble metal electrodes, which are either deposited on a glass or a ceramic substrate. The techniques of thick film printing and thin film deposition are used for depositing these noble metal electrodes. Interdigital electrodes configuration is most widely used. Thin films for humidity sensing are deposited in between the inter digital electrodes [74–77].

Resistive sensors are based on the measurement of change of humidity levels in terms of change in the electrical impedance of the moisture containing medium. The resistive sensors are based on the principle of adsorption of water molecules and their subsequent dissociation into ionic hydroxyl groups. The response time of these type of sensors is quite small(~10s) [78]. The planar thick/thin film humidity sensor based on interdigital structure is shown in the **Figure 6.**

4. Polymer based resistive humidity sensors

These sensors are based on the thin films of porous polymers [79]. The polymeric films take up water vapour molecules from the atmosphere. These molecules condense in the presence of pores of the capillaries. This water vapour intake produces changes in physical or electrical properties of the polymeric material [80, 81].

A new type of humidity sensors has emerged lately based upon polymer electrolytes, which are also called poly electrolytes. Poly electrolytes are a group of polymers, which have electrolytic groups in them and display conductivity, when they are exposed to water vapour molecules. The poly electrolyes can be classified on the basis of their functional electrolytic groups, into following three major categories [29]:

    i. Quaternary ammonium salts

    ii. Sulfonate salts

    iii. Phosphonium salts

Polyelectrolytes are hydrophilic in nature and tend to dissolve in water [82]. But polyelectrolytes based humidity sensors become non-reliable, if the humidity levels in the surroundings is quite high. In comparison to polyelectrolytes, humidity sensors based on conducting polymers like poly (3,4-ethylenedioxythiophene) (PEDOT) or poly (3.4-ethylenedioxythiophene-poly(styrene-sulfonate)(PEDOT-PSS) show high sensitivity to moisture and are partially hydrophobic [83, 84]. Much work has been done by the research investigators in the recent past, to improve the design of humidity sensors based on polymer electrolytes, to make them water resistant. These methods include grafting, copolymerization, cross links formation, interpenetrating network structures, metal oxide added polymers, photochemical cross linking reactions, anchoring of polymer membranes on the electrode surfaces by means of ultraviolet radiation and addition of dopants [85–88].

The creation of strong conjuctions between polymers and plastic substrates is also a subject of research investigations for the last few years. With the introduction of bonding matrices, prepared by different physical and chemical techniques, this problem has been solved to a considerable extent. Electrospinning method has been used to fabricate composite nanofibres, containing silicon-containing polymer electrolyte, polyethylene oxide and polyaniline. It was found that the presence of polyaniline (PANI) in the nanofibres led to the decrease of the impedance of the thin films. It was further observed that the adhesion of film to both the substrate and electrode was due to the formation of nanostructure beads in the nanofibres [89, 90]. To increase the conductance change of resistive polymeric humidity sensors based on conjugated polymers, doping agents are used. Poly (p-diethylnylbenzene) (PDEB) has been synthesised with a nickel catalyst in dioxane toluene mixed solvent system at room temperature for humidity sensing applications [91]. Gold nanoparticles are also used in place of Nickel nanoparticles, to increase the conductivity of the thin films. Polyelectrolytes based resistive humidity sensors also suffer from poor conductivity at low humidity levels. This problem is solved by changing the polymer matrix using a superconductor with high conductivity and also mixing techniques.

5. Ceramic based resistive humidity sensors

New humidity detection mechanisms have been developed to resolve the low sensitivity and selectivity problems of humidity sensors. Porous ceramic humidity sensors are fabricated by a variety of techniques such as thick film screen printing, plasma or vapour deposition. In these kind of thick films, the thickness is always kept more than 10 micrometres and dopants are added to increase the dissociation of water molecules. Thin films prepared by vacuum deposition or plasma deposition act as resistive type devices. The released hydroxyl groups change the impedance of thin film elements as they decrease the resistivity of the thin films. For example, $MgCr_2O_4$-TiO_2 material functioned on the basis of physisorption and chemisorption of water molecules followed by protonic conduction [92]. The humidity sensor elements based on these material showed good conductivity at both low as well as higher relative humidity levels. In this type of humidity sensor, heating was necessary to eliminate the hydroxyl groups on the surface and also to remove contaminants like dust, oil and other types of foreign particles. The ceramic humidity sensor is shown in the **Figure 7** [93].

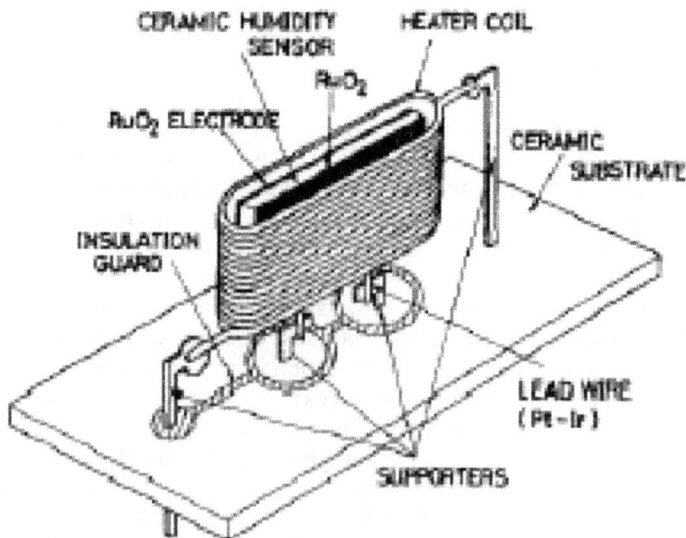

Figure 7.
The above diagram shows a ceramic humidity sensor based on $MgCr_2O_4$-TiO_2.

6. Conclusions/summary

The humidity sensing properties, manufacturing technologies and operating mechanisms of various humidity sensors consisting of different types of materials has been described. Similarly, synthesis and preparation methods for sensors for hygrometric applications have also been explained in this chapter. As protonic conduction type is the most widely accepted mechanism in the majority of humidity sensors, so it is discussed in a detailed manner.

Among all humidity sensor design configurations, the impedance- (resistive) and capacitive-based sensors are the best suited and most popular in the research and industrial environments. Thick and thin film based humidity sensors are also widely used because of cost-effectiveness and ease of fabrication. The humidity sensors based upon ceramic and polymer materials are also used but in lesser magnitude as compared to their above mentioned counterparts and in selected areas of application because of their obvious limitations.

Author details

Rajesh Kumar
Lovely Professional University, Phagwara, Punjab, India

*Address all correspondence to: rajesh.12236@lpu.co.in

IntechOpen

References

[1] Dai, C.-L.; Liu, M.-C.; Chen, F.-S.; Wu, C.-C.; Chang, M.-W. A Nanowire WO3 Humidity Sensor Integrated with Micro-Heater and Inverting Amplifier Circuit on Chip Manufactured Using CMOS-MEMS Technique. *Sens. Actuators B Chem.* **2007**, *123*, 896-901.

[2] Boltshauser, T.; Schonholzer, M.; Brand, O.; Baltes, H. Resonant Humidity Sensors Using Industrial CMOS-Technology Combined with Postprocessing. *J. Micromech. Microeng.* **1992**, *2*, 205-207.

[3] Okcan, B.; Akin, T. A Low-Power Robust Humidity Sensor in a Standard CMOS Process. *IEEE Trans. Electron Devices* **2007**, *54*, 3071-3078.

[4] Oprea, A.; Courbat, J.; Bârsan, N.; Briand, D.; de Rooij, N.F.; Weimar, U. Temperature, Humidity and Gas Sensors Integrated on Plastic Foil for Low Power Applications. *Sens. Actuators B Chem.* **2009**, *140*, 227-232.

[5] Shi, Y.; Luo, Y.; Zhao, W.; Shang, C.; Wang, Y.; Chen, Y. A Radiosonde Using a Humidity Sensor Array with a Platinum Resistance Heater and Multi-Sensor Data Fusion. *Sensors* **2013**, *13*, 8977-8996.

[6] Christian, S. New Generation of Humidity Sensors. *Sens. Rev.* **2002**, *22*, 300-302.

[7] Mehrabani, S.; Kwong, P.; Gupta, M.; Armani, A.M. Hybrid Microcavity Humidity Sensor. *Appl. Phys. Lett.* **2013**, *102*, 241101.

[8] Wang, Y.-H.; Lee, C.-Y.; Chiang, C.-M. A MEMS-Based Air Flow Sensor with a Free-Standing Micro-Cantilever Structure. *Sensors* **2007**, *7*, 2389-2401.

[9] Wolfbeis, O.; Su, P.-G.; Ho, C.-J.; Sun, Y.-L.; Chen, I.-C. A Micromachined Resistive-Type Humidity Sensor with a Composite Material as Sensitive Film. *Sens. Actuators B Chem.* **2006**, *113*, 837-842.

[10] Hanreich, G.; Nicolics, J.; Mündlein, M.; Hauser, H.; Chabicovsky, R. A New Bonding Technique for Human Skin Humidity Sensors. *Sens. Actuators A Phys.* **2001**, *92*, 364-369.

[11] Mohd Syaifudin, A.R.; Mukhopadhyay, S.C.; Yu, P.L. Modelling and Fabrication of Optimum Structure of Novel Interdigital Sensors for Food Inspection. *Int. J. Numer. Model. Electron. Netw. Devices Fields* **2012**, *25*, 64-81.

[12] Stetter, J.R.; Penrose, W.R.; Yao, S. Sensors, Chemical Sensors, Electrochemical Sensors, and ECS. *J. Electrochem. Soc.* **2003**, *150*, S11.

[13] Carr-Brion, K. *Moisture Sensors in Process Control*; Elsevier Applied Science Publishers: London, UK, 1986.

[14] Salehi, A.; Nikfarjam, A.; Kalantari, D.J. Highly Sensitive Humidity Sensor Using Pd/Porous GaAs Schottky Contact. *IEEE Sens. J.* **2006**, *6*, 1415-1421.

[15] Kim, J.-H.; Hong, S.-M.; Lee, J.-S.; Moon, B.-M.; Kim, K. High Sensitivity Capacitive Humidity Sensor with a Novel Polyimide Design Fabricated by MEMS Technology. In Proceedings of 2009 4th IEEE International Conference on Nano/Micro Engineered and Molecular Systems, Shenzhen, China, 5-8 January 2009; pp. 703-706.

[16] Xu, L.; Wang, R.; Xiao, Q.; Zhang, D.; Liu, Y. Micro Humidity Sensor with High Sensitivity and Quick Response/Recovery Based on ZnO/TiO2 Composite Nanofibers. Chin. Phys. Lett. 2011, 28, doi:10.1088/0256-307X/28/7/070702.

[17] Aziz, F.; Sayyad, M.H.; Sulaiman, K.; Majlis, B.H.; Karimov, K.S.; Ahmad,

Z.; Sugandi, G. Corrigendum: Influence of Humidity Conditions on the Capacitive and Resistive Response of an Al/VOPc/Pt Co-Planar Humidity Sensor. *Meas. Sci. Technol.* **2012**, 23, 069501.

[18] Lin, W.-D.; Chang, H.-M.; Wu, R.-J. Applied Novel Sensing Material Graphene/polypyrrole for Humidity Sensor. Sens. Actuators B Chem. 2013, 181, 326-331.

[19] Yadav, B.C.; Singh, M. Morphological and Humidity Sensing Investigations on Niobium, Neodymium, and Lanthanum Oxides. *IEEE Sens. J.* **2010**, 10, 1759-1766.

[20] Xu, C.-N.; Miyazaki, K.; Watanabe, T. Humidity Sensors Using Manganese Oxides. *Sens. Actuators B Chem.* **1998**, 46, 87-96.

[21] Pelino, M.; Colella, C.; Cantalini, C.; Faccio, M.; Ferri, G.; D'Amico, A. Microstructure and Electrical Properties of an A-Hematite Ceramic Humidity Sensor. *Sens. Actuators B Chem.* **1992**, 7, 464-469.

[22] Klym, H.; Hadzaman, I.; Shpotyuk, O.; Brunner, M. P3.6—Multifunctional T/RH-Sensitive Thick-Film Structures for Environmental Sensors. In Proceedings of SENSOR + TEST Conferences 2011, Nürnberg, Germany, 7-9 June 2011; pp. 744-748.

[23] Traversa, E.; Bearzotti, A. A Novel Humidity-Detection Mechanism for ZnO Dense Pellets. Sens. Actuators B Chem. 1995, 23, 181-186.

[24] Gusmano, G.; Montesperelli, G.; Nunziante, P.; Traversa, E. Study of the Conduction Mechanism of MgAl2O4 at Different Environmental Humidities. Electrochim. Acta 1993, 38, 2617-2621

[25] Morten, B.; Prudenziati, M.; Taroni, A. Thick-Film Technology and Sensors. *Sens. Actuators* **1983**, 4, 237-245.

[26] Smetana, W.; Unger, M. Design and Characterization of a Humidity Sensor Realized in LTCC-Technology. *Microsyst. Technol.* **2007**, 14, 979-987.

[27] Karimov, K.S.; Cheong, K.Y.; Saleem, M.; Murtaza, I.; Farooq, M.; Noor, A.F.M. Ag/PEPC/NiPc/ZnO/Ag Thin Film Capacitive and Resistive Humidity Sensors. *J. Semicond.* **2010**, 31, 054002.

[28] Fraden, J. *Handbook of Modern Sensors*; Springer New York: New York, NY, 2010; pp. 445-459.

[29] *HIH Series Humidity Sensors*; Psychrometrics and Moisture, Reference and Application Data. Honeywell: Morristown, NJ, USA; pp. 145-147.

[30] Pokhrel, S.; Jeyaraj, B.; Nagaraja, K.S. Humidity-Sensing Properties of ZnCr2O4-ZnO Composites. *Mater. Lett.* **2003**, 22-23, 3543-3548.

[31] Chen, Y.S.; Li, Y.; Yang, M.J. Humidity Sensitive Properties of NaPSS/MWNTs Nanocomposites. *J. Mater. Sci.* **2005**, , 5037-5039.

[32] Zhang, Y.; Yu, K.; Jiang, D.; Zhu, Z.; Geng, H.; Luo, L. Zinc Oxide Nanorod and Nanowire for Humidity Sensor. Appl. Surf. Sci. 2005, 242, 212-217.

[33] Kuang, Q.; Lao, C.; Wang, Z.L.; Xie, Z.; Zheng, L. High-Sensitivity Humidity Sensor Based on a Single SnO2 Nanowire. *J. Am. Chem. Soc.* **2007**, 129, 6070-6071.

[34] Kassas, A. Humidity Sensitive Characteristics of Porous Li-Mg-Ti-O-F Ceramic Materials. *Am. J. Anal. Chem.* **2013**, 04, 83-89.

[35] Wang, K.; Qian, X.; Zhang, L.; Li, Y.; Liu, H. Inorganic-Organic P-N Heterojunction Nanotree Arrays for a High-Sensitivity Diode Humidity Sensor. *ACS Appl. Mater. Interfaces* **2013**, 5, 5825-5831.

[36] Kulwicki, B. Humidity Sensors. *J. Am. Ceram. Soc.* **1991**, *74*, 697-708.

[37] Traversa, E. Ceramic Sensors for Humidity Detection: The State-of-the-Art and Future Developments. *Sens. Actuators B Chem.* **1995**, *23*, 135-156.

[38] Chen, Z.; Lu, C. Humidity Sensors: A Review of Materials and Mechanisms. *Sens. Lett.* **2005**, *3*, 274-295.

[39] Salehi, A.; Kalantari, D.J.; Goshtasbi, A. Rapid Response of Au/Porous-GaAs Humidity Sensor at Room Temperature. In Proceedings of 2006 Conference on Optoelectronic and Microelectronic Materials and Devices, Perth, Australia, 6-8 December 2006; pp. 125-128.

[40] Shah, J.; Kotnala, R.K.; Singh, B.; Kishan, H. Microstructure-Dependent Humidity Sensitivity of Porous MgFe2O4-CeO2 Ceramic. *Sens. Actuators B Chem.* **2007**, *128*, 306-311.

[41] Dunmore, F. An Electric Hygrometer and Its Application to Radio Meteorography. *J. Res. Natl. Bur. Stand.* **1938**, *20*, 723-744.

[42] Packirisamy, M.; Stiharu, I.; Li, X.; Rinaldi, G. A Polyimide Based Resistive Humidity Sensor. Sens. Rev. 2005, 25, 271-276.

[43] Cho, N.-B.; Lim, T.-H.; Jeon, Y.-M.; Gong, M.-S. Inkjet Printing of Polymeric Resistance Humidity Sensor Using UV-Curable Electrolyte Inks. *Macromol. Res.* **2008**, *16*, 149-154.

[44] Matsuguchi, M. A Capacitive-Type Humidity Sensor Using Cross-Linked Poly(methyl Methacrylate) Thin Films. J. Electrochem. Soc. 1991, 138, 1862.

[45] Mukode, S.; Futata, H. Semiconductive Humidity Sensor. *Sens. Actuators* **1989**, *16*, 1-11.

[46] Shimizu, Y. Humidity-Sensitive Characteristics of La3+-Doped and Undoped SrSnO3. *J. Electrochem. Soc.* **1989**, *136*, 1206

[47] Tulliani, J.-M.; Baroni, C.; Zavattaro, L.; Grignani, C. Strontium-Doped Hematite as a Possible Humidity Sensing Material for Soil Water Content Determination. *Sensors* **2013**, *13*, 12070-12092.

[48] Shimizu, Y.; Arai, H.; Seiyama, T. Theoretical Studies on the Impedance-Humidity Characteristics of Ceramic Humidity Sensors. Sens. Actuators 1985, 7, 11-22.

[49] Seiyama, T.; Yamazoe, N.; Arai, H. Ceramic Humidity Sensors. *Sens. Actuators* **1983**, *4*, 85-96.

[50] Fripiat, J.J.; Jelli, A.; Poncelet, G.; André, J. Thermodynamic Properties of Adsorbed Water Molecules and Electrical Conduction in Montmorillonites and Silicas. *J. Phys. Chem.* **1965**, *69*, 2185-2197.

[51] McCafferty, E.; Pravdic, V.; Zettlemoyer, A.C. Dielectric Behaviour of Adsorbed Water Films on the A-Fe2O3 Surface. *Trans. Faraday Soc.* **1970**, *66*, 1720.

[52] Bernal, J.D.; Fowler, R.H. A Theory of Water and Ionic Solution, with Particular Reference to Hydrogen and Hydroxyl Ions. *J. Chem. Phys.* **1933**, *1*, 515.

[53] Agmon, N. The Grotthuss Mechanism. Chem. Phys. Lett. 1995, 244, 456-462.

[54] Wraight, C.A. Chance and Design—Proton Transfer in Water, Channels and Bioenergetic Proteins. *Biochim. Biophys. Acta* **2006**, *1757*, 886-912.

[55] Conway, B.E.; Bockris, J.O.; Linton, H. Proton Conductance and the Existence of the H3O Ion. *J. Chem. Phys.* **1956**, *24*, 834.

[56] Yates, D.J. C. Infrared Studies of the Surface Hydroxyl Groups on Titanium Dioxide, and of the Chemisorption of Carbon Monoxide and Carbon Dioxide. *J. Phys. Chem.* **1961**, *65*, 746-753.

[57] Blyholder, G.; Richardson, E.A. Infrared and volumetric data on the adsorption of ammonia, water, and other gases on activated iron(iii) oxide 1. *J. Phys. Chem.* **1962**, *66*, 2597-2602.

[58] Young, G. Interaction of Water Vapor with Silica Surfaces. J. Colloid Sci. 1958, 13, 67-85.

[59] Colomban, P. *Proton Conductors: Solids, Membranes and Gels—Materials and Devices*; Cambridge University Press: Cambridge, UK, 1992; pp. 581.

[60] Anderson, J.H.; Parks, G.A. Electrical Conductivity of Silica Gel in the Presence of Adsorbed Water. J. Phys. Chem. 1968, 72, 3662-3668.

[61] Morimoto, T.; Nagao, M.; Tokuda, F. Relation between the Amounts of Chemisorbed and Physisorbed Water on Metal Oxides. *J. Phys. Chem.* **1969**, *73*, 243-248.

[62] Hair, M.L.; Hertl, W. Adsorption on Hydroxylated Silica Surfaces. J. Phys. Chem. 1969, 73, 4269-4276.

[63] McCafferty, E.; Zettlemoyer, A.C. Adsorption of Water Vapour on α-Fe2O3. *Discuss. Faraday Soc.* **1971**, *52*, 239

[64] Hertl, W.; Hair, M.L. Hydrogen Bonding between Adsorbed Gases and Surface Hydroxyl Groups on Silica. *J. Phys. Chem.* **1968**, *72*, 4676-4682.

[65] Thiel, P.A.; Madey, T.E. The Interaction of Water with Solid Surfaces: Fundamental Aspects. *Surf. Sci. Rep.* **1987**, *7*, 211-385.

[66] Nitta, T.; Hayakawa, S. Ceramic Humidity Sensors. *IEEE Trans.*

Components Hybrids Manuf. Technol. **1980**, *3*, 237-243.

[67] Fripiat, J.J.; Uytterhoeven, J. Hydroxyl Content in Silica Gel — Aerosil‖. *J. Phys. Chem.* **1962**, *66*, 800-805.

[68] Fagan, J.G.; Amarakoon, R.W. Reliability and Reproducibility of Ceramic Sensors. III: Humidity Sensors. Am. Ceram. Soc. Bull. 1993, 72, 119-130.

[69] Kurosaki, S. The Dielectric Behavior of Sorbed Water on Silica Gel. *J. Phys. Chem.* **1954**, *58*, 320-324.

[70] Gusmano, G.; Nunziante, P.; Traversa, E.; Montanari, R. Microstructural Characterization of MgFe2O4 Powders. *Mater. Chem. Phys.* **1990**, *26*, 513-526.

[71] Gusmano, G.; Montesperelli, G.; Traversa, E.; Mattogno, G. Microstructure and Electrical Properties of MgAl2O4 Thin Films for Humidity Sensing. *J. Am. Ceram. Soc.* **1993**, *76*, 743-750.

[72] Foster, A.G. The Sorption of Condensible Vapours by Porous Solids. Part I. The Applicability of the Capillary Theory. *Trans. Faraday Soc.* **1932**, *28*, 645.

[73] Yamazoe, N.; Shimizu, Y. Humidity Sensors: Principles and Applications. *Sens. Actuators* **1986**, *10*, 379-398.

[74] Traversa, E.; Sadaoka, Y.; Carotta, M.C.; Martinelli, G. Environmental Monitoring Field Tests Using Screen-Printed Thick-Film Sensors Based on Semiconducting Oxides. *Sens. Actuators B Chem.* **2000**, *65*, 181-185.

[75] Kunte, G.V.; Shivashankar, S.A.; Umarji, A.M. Humidity Sensing Characteristics of Hydrotungstite Thin Films. *Bull. Mater. Sci.* **2009**, *31*, 835-839.

[76] Mamishev, A.V.; Sundara-Rajan, K.; Zahn, M. Interdigital Sensors and

Transducers. *Proc. IEEE* **2004**, *92*, 808-845.

[77] Moneyron, J.E.; de Roy, A.; Besse, J.P. Realisation of a Humidity Sensor Based on the Protonic Conductor Zn2Al(OH)6Cl.nH2O. *Microelectron. Int.* **1991**, *8*, 26-31.

[78] Sakai, Y.; Sadaoka, Y.; Matsuguchi, M. Humidity Sensors Based on Polymer Thin Films. *Sens. Actuators B Chem.* **1996**, *35*, 85-90.

[79] Harris, K.D.; Huizinga, A.; Brett, M.J. High-Speed Porous Thin Film Humidity Sensors. *Electrochem. Solid-State Lett.* **2002**, *5*, H27.

[80] Tsuchitani, S.; Sugawara, T.; Kinjo, N.; Ohara, S.; Tsunoda, T. A Humidity Sensor Using Ionic Copolymer and Its Application to a Humidity-Temperature Sensor Module. *Sens. Actuators* **1988**, *15*, 375-386.

[81] Rauen, K.L.; Smith, D.A.; Heineman, W.R.; Johnson, J.; Seguin, R.; Stoughton, P. Humidity Sensor Based on Conductivity Measurements of a Poly(dimethyldiallylammonium Chloride) Polymer Film. *Sens. Actuators B Chem.* **1993**, *17*, 61-68.

[82] Li, Y. A Novel Highly Reversible Humidity Sensor Based on poly(2-Propyn-2-Furoate). *Sens. Actuators B Chem.* **2002**, *86*, 155-159.

[83] Ogura, K.; Saino, T.; Nakayama, M.; Shiigi, H. The Humidity Dependence of the Electrical Conductivity of a Solublepolyaniline—Poly(vinyl Alcohol) Composite Film. *J. Mater. Chem.* **1997**, *7*, 2363-2366.

[84] Liu, J.; Agarwal, M.; Varahramyan, K.; Berney, E.S.; Hodo, W.D. Polymer-Based Microsensor for Soil Moisture Measurement. Sens. Actuators B Chem. 2008, 129, 599-604.

[85] Lee, C.-W.; Park, H.-S.; Gong, M.-S. Humidity-Sensitive Properties of

Polyelectrolytes Containing Alkoxysilane Crosslinkers. *Macromol. Res.* **2004**, *12*, 311-315.

[86] Sakai, Y.; Sadaoka, Y.; Ikeuchi, K. Humidity Sensors Composed of Grafted Copolymers. *Sens. Actuators* **1986**, *9*, 125-131.

[87] Sakai, Y.; Sadaoka, Y.; Matsuguchi, M.; Rao, V.L.; Kamigaki, M. A Humidity Sensor Using Graft Copolymer with Polyelectrolyte Branches. Polymer 1989, 30, 1068-1071.

[88] Moreno-Bondi, M.C.; Orellana, G.; Li, Y.; Yang, M.J.; She, Y. Humidity Sensitive Properties of Crosslinked and Quaternized poly(4-Vinylpyridine-Co-Butyl Methacrylate). *Sens. Actuators B Chem.* **2005**, *107*, 252-257.

[89] Li, P.; Li, Y.; Ying, B.; Yang, M. Electrospun Nanofibers of Polymer Composite as a Promising Humidity Sensitive Material. *Sens. Actuators B Chem.* **2009**, *141*, 390-395.

[90] Su,P.-G.; Hsu, H.-C.; Liu, C.-Y. Layer-by-Layer Anchoring of Copolymer of Methyl Methacrylate and [3-(methacrylamino)propyl] Trimethyl Ammonium Chloride to Gold Surface on Flexible Substrate for Sensing Humidity. *Sens. Actuators B Chem.* **2013**, *178*, 289-295.

[91] Yang, M.; Li, Y.; Zhan, X.; Ling, M. A Novel Resistive-Type Humidity Sensor Based on Poly(p-Diethynylbenzene). *J. Appl. Polym. Sci.* **1999**, *74*, 2010-2015.

[92] NITTA, T.; TERADA, Z.; HAYAKAWA, S. Humidity-Sensitive Electrical Conduction of MgCr2O4-TiO2 Porous Ceramics. *J. Am. Ceram. Soc.* **1980**, *63*, 295-300.

[93] Nitta, T.; Hayakawa, S. Ceramic Humidity Sensors. *IEEE Trans. Components Hybrids Manuf. Technol.* **1980**, *3*, 237-243.

Chapter 3

Humidity Sensors, Major Types and Applications

Jude Iloabuchi Obianyo

Abstract

The need for humidity sensors in various fields have led to the development and fabrication of sensors for use in industries such as the medical, textile, and laboratories. This chapter reviewed humidity sensors, major types and applications with emphasis on the optical fiber, nanobricks, capacitive, resistive, piezoresistive and magnetoelastic humidity sensors. While optical fiber sensors are best for use in harsh weather conditions, the nanobricks sensors have excellent qualities in humidity sensing. Capacitive sensors make use of impedance and are more durable than the equivalent resistive sensors fabricated with ceramic or organic polymer materials and have short response and recovery times which attest to their efficiency. Piezoresistive sensors have fast response time, highly sensitive and can detect target material up to one pictogram range. Magnetoelastic sensors are very good and can measure moisture, temperature and humidity between 5% and 95% relative humidity range. It was concluded that sensors have peculiar applications.

Keywords: Humidity sensors, Major types, Applications, A Review

1. Introduction

1.1 Humidity

Humidity is a measure of quantity of water vapor present in the air or a gas. Humidity is a general terminology and is used to measure the amount of water vapor in any given environment. Measurement of this parameter is important in the environment because many appliances in the industries cannot be stored under certain range of humidity. High humidity entails high atmospheric moisture content, hence condensation which depends on the atmospheric temperature resulting in corrosion of metals and similar features in industries.

The industries that could be affected include hospitals, textile industries, laboratories, storage rooms for computers, food processing industries, art museums, shopping malls, libraries, exhibition centers, pharmaceutical stores. For these reasons, humidity sensors are widely used in numerous fields of human endeavor such as for weather forecasting, food processing, health care etc. It is good practice to measure humidity from time to time for maintenance of optimal environmental conditions suitable for products in order to prevent damage.

2. Humidity sensors

Humidity is measured with a device known as humidity sensor and different types are available in the market such as the, optical, gravimetric, capacitive, resistive, piezoresistive and magnoelastic sensors etc. Each of these humidity sensors have peculiar applications which depend on the design and suitability for a given environment. Humidity sensors are electronic equipments that can be used to measure humidity in any given environment. Being electronic devices, they transform information to equivalent electrical signal. Humidity sensors have very wide applications, functionality and appear in different sizes. Sometimes, they are incorporated in handheld devices as in smart phones.

2.1 Optical fiber humidity sensors

Being a recent technology in sensor development, optical fiber sensors have peculiar advantages such as lightweight, chemical stability, handy, non-susceptible to electromagnetic fields and the ability to communicate two or more signals over a common channel [1]. Because of these advantages, optical fiber sensors essentially is a dependable technology that can be applied in many fields such as industries, medicine, and structural health monitoring [2–4]. Apart from its excellent performance in measurement of humidity, Its application can be extended to the measurements of angle, refractive index, temperature, acceleration, pressure, breathing rate and oxygen saturation [5–10].

Optical sensors exhibit optimal performance in harsh environmental conditions. Measurement of relative humidity, influence of coating thicknesses and temperature with the use of polyimide-recoated optical fiber Bragg gratings showed that relative humidity (RH) and temperature (T) sensitivities were $S_{RH} = (2.21 \pm 0.10) \times 10^{-6} \%RH^{-1}$ and $S_T = (7.79 \pm 0.08) \times 10^{-6} K^{-1}$, these indicate a RH sensitivity of 44.2% and a T sensitivity of 86.6% [11]. Apart from the use of optical fiber sensors in measuring humidity, it can equally be used as a temperature sensing device because of its high sensitivity of $1.04 \times 10^{-3} °C^{-1}$ with a linearity of 0.994, this device has root mean square of $1.48°C$ an indication of 2% relative error [7].

Embedment of optic fiber sensors in metallic materials such as nickel- and copper-coated fiber Bragg grating (FBG) shields sensors from environmental influence and gives rise to better performance. Though, Copper-coated sensors lost their sensitivity to temperature and strain in the process of embedment using tungsten inert gas (TIG) welding due to thermal and mechanical strain, while nickel-coated sensors were less sensitive to the same effects and showed better sensor performance [12].

Attachment of fiber optic sensors on metallic structures have proved to be important in order to enhance the durability of sensors since non-embedded sensors are prone to damage. Laser cladding technology was used as an alternative in embedding metal coated fiber optics in which fiber Bragg grating is incorporated. This device minimized the high thermal and mechanical strain usually generated with TIG welding. It was shown that this sensor gave room for satisfactory results for strain and temperature measurements, and was favorably compared with the conventional gauge used in calibration of sensors [13].

Design is an important factor in performance of humidity sensors since different materials exhibit different magnitudes of sensitivity to humid environment. Performance of a humidity sensor is measured by its detection range. Magnesium oxide (MgO) humidity sensor designed with micro-arc oxidation (MAO) technology gave better performance than semiconductor humidity sensors which are usually prone to narrow detection ranges and poor sensitivities for detection. It has the advantage of using both impedance and capacitance as response signals, giving an output of 150 in the low relative humidity (RH) range (11.3–67% RH) when impedance was the response signal, and an output 120 at high humidity range of (67–97.3% RH) with capacitance as the response signal [14].

2.2 Nanobricks humidity sensors

Mesoporous tungsten oxide nanobricks (WO_3 NBs) have proved to be excellent sensors because of its sensitivity to ammonia gas, volatile organic compounds and humid environment at room temperature [15]. It has a 75% response which is on the high side, a 15-day operational stability at 100 ppm ammonia concentration and extremely-high response/recovery time of 8/5 s. Generally, this family of sensors exhibit 32% resistance response at 20% RH with fast response/recovery time of 10/8 s, high specific surface area of the monoclinic crystal structure is instrumental to the high sensitivity of this device.

Nanobricks sensors such as Ag-CuO/rGO (5 at % Ag) have shown maximum response of approximately 52% RH humidity when operated at approximately 22°C. This sensor made of pure Ag-CuO nanostructures containing varied quantities of Ag produced by hypothermal method is highly sensitive to the presence of NO_2 between a temperature range of 22–100°C [16].

Carbon nanotubes invented by [17] have extraordinary electrical and mechanical properties and for this reason are found in a number of applications [18, 19]. Due to their notable electrical characteristic, many researchers have studied the application of this material as a sensor not only because of its eminent electrical feature, but the reduced costs of manufacture which makes them very attractive for use by sensing industries [20]. Suspended aligned nanotubes beams was generated to asses its humidity sensing characteristics. The device was produced at room temperature using epoxy-based highly functional photoresist referred as SU8, and for 15–98% RH range, [20] reported a threefold better performance for suspended devices which exhibited improved sensitivity and little hysteresis at 10 humidity cycles when compared with non-suspended device. This implies that suspension of carbon nanotubes beam improved the sensitivity of the sensor, and the sensitivity factor is given by the expression;

$$S = \frac{R_H - R_0}{R_0} \times 100 \qquad (1)$$

where R_H is the resistance at the measured value of resistance, and R_0 is value of baseline resistance.

Zinc stannate ($ZnSnO_3$) nanoparticles prepared by chemical precipitation have shown to be good as humidity sensing device. The X-ray power diffraction of this sensor presented $ZnSnO_3$ as having a perovskite phase with orthorhombic structure and a minimum of 4 nm crystalline size. Comparative study of $ZnSnO_3$ pellet annealed at 600°C for 1:4 weight ratio and pure SnO_3 by [21] indicate that $ZnSnO_3$ is more sensitive to humidity than SnO_3 when subjected to the same environmental conditions. They concluded that zinc stannate nanoparticles showed a maximum sensitivity of 3 GΩ/% RH when compared with the equivalent SnO_3 used in this study.

In another research, it has been reported by [22] that synthesized Fe_3O_4-Nps with porous structure has shown exceptional qualities in humidity sensing. Characterization of prepared Fe_3O_4-Nps was done with the aid of X-ray diffraction, (XRD), transmission electron microscope (TEM) and vibration sample magnetometer (VSM), which revealed the spherical pores structure. With this, magnetic nanostructures (MNs) are highly valued by researchers and humidity sensors industries because they posses vital characteristics that make them to be widely applied.

2.3 Capacitive humidity sensors

Organic materials such as hydrophobic polymers are employed in the manufacture of capacitive humidity sensors. Capacitive humidity sensors are more durable than the equivalent resistive humidity sensors because it has the propensity for heavy water vapor condensation at high humidity level [23], and may operate comfortably at relatively higher temperature of about 200°C [24], with capacitance range of between 100 to 500 pF at intervals of 50% changes in humidity level [25]. Capacitive sensors are easily available in the market because of their better linearity characteristics when compared with the resistive type sensors [26, 27]. Capacitive thin-film humidity sensors are responsive to changes in RH because their sensors rely on the dielectric constant value of the active layers. Response time of capacity sensors is a function of three parameters namely the sensor design, materials propensity for sorption and desorption of water vapor, and the sensor temperature [28].

The most important thing in capacitive humidity sensors is the material it is made of. The hygroscopic characteristics of the hydrophobic elements in these sensors are instrumental to its ability to absorb water molecules from their surrounding and as a result, this should be born in mind during construction so as to produce a good configuration that would optimally utilize these properties for effective performance [29]. Three major factors affect the performance of capacitive humidity sensors and include; surface area of the electrode, dielectric material polarization and distance apart between the electrodes.

As earlier stated that humidity sensors have vary wide applications. In agriculture, there is need to measure the moisture/humidity levels to guide in decision making regarding the optimal ambient humidity condition that will be suitable for excellent crop yield.

Without doubt, water is very important in the physicochemical and mechanical characteristics of soil. Variations in soil moisture quantity can have significant influence on ecosystems, plant growth and biodiversity. Capacitive moisture sensors are used for sensing the moisture contents of soils in agricultural endeavors. Availability of soil moisture in the root zone is very crucial for plants development and growth which depends on the physical characteristics of soil and the surrounding environment with respect to climatological conditions [30].

The output of a capacitive moisture sensor is a function of the complex relative permittivity ε_r^* of the soil which is the dielectric medium given by the expression [31];

$$\varepsilon_r^* = \varepsilon_r' - j\varepsilon_r'' = \varepsilon_r' - j\left(\varepsilon_{relax}' + \frac{\sigma_{dc}}{2\pi f \varepsilon_0}\right), \tag{2}$$

where ε_r' and ε_r'' are the real and the imaginary part of the permittivity respectively, σ_{dc} is the electrical conductivity, ε_{relax}'', is the molecular relaxation contribution (dipolar rotational, atomic vibrational, and electronic energy states, j is the imaginary number $\sqrt{-1}$, and f the frequency. However, it has been reported that capacitive coplanar soil moisture sensor showed reliable relationship between output voltage and gravimetric water content [32].

Another type of humidity sensor are the ones made with two-dimensional materials. The two-dimensional (2D) materials show superior physical characteristics when compared with the corresponding one-directional materials on application of charge and heat to a planar layer [33]. The two dimensional nature of a material give a higher ratio of surface area to volume and hence, makes it easier for fabrication of sensing layers.

Some of these humidity sensors can neither be fully classified as belonging to either the capacitive or resistive humidity sensors because they make use of combined effects of capacitance and impedance in sensing the humidity of environment. Example of this is the humidity sensor based on two dimensional molybdenium diselenide ($MoSe_2$). Material for this sensor is the transition metal dichalcogenides (TMDCs) which are very good in fabrication of sensors for applications in industries. This humidity sensor has been reported to be highly stable and effectively superfast in detecting the moisture level in its surroundings. At frequency of 1 kHz and between the temperature range of 20–30°C, impedance witnessed a hysteresis of less than 1.98% while the capacitance had a hysteresis of less than 2.36%, maximum error rates of −0.162% and −0.183% were observed based on the impedance and capacitance responses respectively recorded between humidity range of 0–90% RH. Impedance response ((t_{res}) and recovery times (t_{rec}) for the sensor are −0.96 s and −1.03 s. The corresponding response and recovery times for capacitance are −1.87 s and 2.13 s respectively [34].

The conduction mechanism of the sensor is such that as $MoSe_2$ nano-flakes responds to humidity, dielectric constant increases relative to the dry film, there exists a corresponding ionic flow through the sensor. When the sensor is exposed to humidity which invariably comes in contact with the hydroxyl ions, the thin films of $MoSe_2$ tend to absorb water molecules thus establishing ionic conduction paths between $MoSe_2$ nanoflakes, thereby reducing the overall sheet resistance.

In 2004, mechanically exfoliated 2D graphene was discovered which is a new innovation in nanotechnology with several advantages such as transparency, flexibility and high carrier mobility. Major limitations in applicability of 2D graphene is the virtually zero band gap of this material which consequently leads to obstruction of its need for fabrication of sensors due to low on/off ratio and long response and recovery times of approximately 10 s [35–39]. TMDCs is a two-dimensional semiconductor, less expensive and very fast in detection of humidity signals and has a high carrier mobility of about 500 cm^2/Vs.

Apart from its high applicability in humidity sensing because of its exceptional physical properties and high carrier mobility, they are also used in gas, temperature, electronic [40–47] and optoelectronics sensors [48–52].

2.4 Resistive humidity sensors

The working principle of resistive humidity sensors is the ability for the sensor to detect vapor in its surrounding which has direct influence on the electrical resistance of the sensing layer [53] In this sensor device, humidity increase give rise to increase in electrical conductivity which invariably lowers the system resistivity within 1 to 100 kΩ [54]. This sensor has subdivisions such as the electronic and ionic conduction types and are based on the mechanism by which the signals are received by the sensing material. The electronic type is made up of polyelectrolytes which respond to changes in water vapor in the surrounding by affecting the resistivity. The ionic conduction type depend on changes in the dielectric constant of the polymer dielectrics which gives rise to the two categories of resistive type sensors namely; the ionic-conduction [55, 56] and the electronic-conduction [57, 58] respectively.

Resistive humidity sensors can be fabricated with either ceramic or organic polymer materials. However, the ceramic type have shown to be advantageous over the organic polymer type evidenced from the performance of equivalent sensor fabricated with metal oxides materials [54]. Two methods are available for preparation of ceramic metal-oxides for applications in humidity sensors namely, the conventional method [59] and the advanced method [60], meant to impart porosity [61–63] to the device, a characteristic which enhances the efficiency of the sensor.

Innovations and application of internet of things have improved the characteristics of these humidity sensors. These characteristics include low power consumption, room-temperature operation, small size and compatibility with other platforms such as thermal and chemical compatibility during fabrication, easy selection from other sensors for use in calibration of the newly fabricated sensor [64–68].

For instance, resistive and capacitive humidity sensor was fabricated using the new bis(4-benzylpiperazine-1-carbodithioato-k2S,S′)nickel(II) complex, with Von Grotthuss as the conduction mechanism. Between humidity range of 30–90%, [69] reported that the resistance of the sensor decreased by two orders of magnitude. The result showed that at 30%RH, the resistance was $2.94 \times 10^8 \Omega$ and at 90%RH, it was found to be $2.34 \times 10^6 \Omega$ giving a range of 2.92×10^8. A hysteresis of 1.54% was observed with response and recovery times of 25 and 30 seconds. Response and recovery times of approximately 0.14s and 1.7s between a humidity range of 0–70% were reported by [70] after using resistive humidity sensor comprised of biopolymer-derived carbon thin film and carbon microelectrodes, evidencing the super-sensitivity of this sensor when compared with the bis(4-benzylpiperazine-1-carbodithioato-k2S,S′)nickel(II) complex. These short response and recovery times attest to the efficient performance of this sensor and is attributed to the shellac-derived carbon (SDC) film which enables sharp absorption and desorption equilibrium. This sensor is based on a shellac-derived carbon (SDC) active film deposited on sub-micrometer-sized carbon interdigitated electrodes (cIDEs) which is responsible for the optimization of the response and recovery times respectively. Characterization of this SDC-cIDEs-based humidity sensor revealed excellent dynamic range of between 0–90% RH, with a dynamic response of 50% and very high sensitivity of 0.54/% RH.

Nitrogen-doped layers incorporated in humidity sensors are unique and have proved to be excellent humidity sensor device. Example of this sensor is the configuration of nitrogen-doped reduced graphene oxide (nRGO) placed on a colourless polyimide film. This sensor has a detection range of 6.1% to 66.4% RH, the results also showed that a 1.36-fold better performance in terms of sensitivity is achieved when platinum nanoparticles are attached on the surface of nRGO as compared with the pure nRGO. These sensitivities are in the neighbourhood of 4.51% for the Pt-nRGO at 66.4% RH and 3.53% for the nRGO at 66.4% RH [71].

2.5 Piezoresistive humidity sensors

Cantilever-based nanomechanical sensors have two important qualities, fast response time, highly sensitive and real time and label-free detection ability. These important features have made these sensors to be very useful instruments in advanced detection of molecules and can be widely applied in numerous fields [72–75]. These sensors can be used in process monitoring, gas sensors, in hospitals for diagnostic biosensing, and in detection of solvent vapours. One significant feature of this sensor is that they make use of poly-coated cantilever as their active layers [76, 77]. These sensors function as a result of the induced stress from the adsorption of molecules on the sensing layer. The advantage of nanomechanical sensors are their ability to detect target materials up to one trillionth of a gram range, but one major limitation to its application is the bulky size of the optical measurement system [78] that uses a piezoresistive sensor. The self-sensing method approach is usually employed to overcome this shortcoming [79].

Highly miniaturized nanomechanical SI-polymer composite membrane-type internal-stress sensor (MIS) with piezoresistive elements have shown to be an excellent device for humidity detection. This sensor has a surface area of 500 μm^2 and consists of a thin SI-polymer composite membrane supported by two piezoresistive beams. It has a relative resistance of 0.6% at a relative humidity of 58% and a response of 5.2 mV/% RH to 70% relative humidity range, a sensing resolution of 0.5% humidity and polymer expansion ratio, ϵ_p of 2.4×10^{-3}. A perfectly correlated linear relationship existed between the sensor output and relative humidity at 19.5°C, while the Fast Fourier Transform (FFT) analysis of the sensor system for noise resolution at 60% RH was 2.5 mV/$\sqrt{Hz}$ [79].

2.6 Magnetoelastic humidity sensors

Uncrystallized alloys made up of Fe, Ni and Co have proved to be excellent magnetoelastic materials and are very good in fabrication of magnetoelastic humidity sensors. These materials are prone to change of shape on exposure to magnetic fields so that on application of mechanical stress, result in magnetization. The dual directional connection is instrumental to the functioning of this sensor device [80–83]. It is designed in such a manner that sample vibration is directly proportional to the frequency of AC magnetic field applied to the sensor system.

Typical example of this sensor is the TiO_2 nanotubes (TiO_2-NTs) coated with $Fe_{40}Ni_{28}Mo_4B_{18}$ amorphous ferromagnetic ribbons as a humidity sensor, can be used to measure moisture values between the range of 5–95%. Measurement precision of this sensor is very high with low hysteresis. Sensor resonance frequency was approximately 3180 Hz which indicated highly significant change when compared with other magnetoelastic humidity sensors [84]. However, the resonance frequency shift of magnetoelastic sensors depend on two parameters expressed as in Eq. (3); [82, 85–87].

$$\Delta f = -\frac{f}{2}\frac{\Delta m}{M} \qquad (3)$$

where Δm = the variation of sensor mass, M = mass of the magnetoelastic sensors (MES) prior to adsorption.

MES have been reported to be very good in the measurement of humidity and temperature [88–92] and mass [93, 94]. Different methods are available for sensing in the environment such as the use of flow cytometry, immunosensors, and gas chromatography in sensing of volatile organic compounds (VOC) though laborious, time consuming and expensive as limitations to its applications. Acoustic wave based sensors are better being highly sensitive in detection of VOCs, bacteria etc., though with very high sensitivities, it is a wireless sensing instrument and too expensive to afford [95, 96].

3. Conclusions

Humidity sensors are very essential in industries because of the roles they play in giving information about the ambient environmental conditions for products storage. A knowledge which is very important in materials management since different products require certain amount of atmospheric moisture in its surroundings beyond or below which deterioration sets in. Therefore this chapter reviewed humidity sensors, its major types and applications in different fields. Emphasis was on the optical fiber, nanobricks, capacitive, resistive, piezoresistive and magnetoelastic humidity sensors respectively. Each of these humidity sensors have peculiar applications because while the optical fiber sensor perform excellently in harsh environmental conditions, the nanobricks are excellent sensors because of its sensitivity to ammonia gas, volatile organic compounds and humid environment at room temperature. While the nanobricks sensors operate more comfortably at room temperature, capacitive sensors can operate comfortably at relatively higher temperature of about 200°C and have better linearity characteristics when compared with the resistive type sensors.

Resistive humidity sensors fabricated with either ceramic or organic polymer materials have shown to be very good in humidity sensing as a result of increase in electrical conductivity that lowers the electrical resistivity with very short response and recovery times. Piezoresistive sensors are equally good in humidity sensing with low relative resistance and a perfectly correlated linear relationship which exist between the sensor output and relative humidity, while the magnetoelastic humidity sensors fabricated from alloys of iron, nickel and cobalt have wonderful performance in humidity sensing because of their high measurement precision and low hysteresis. These humidity sensors do not have universal applications because of their peculiarities since each of them is more suitable for sensing in a given environment and recommendation should be based on this notion.

Author details

Jude Iloabuchi Obianyo
Department of Civil Engineering, Michael Okpara University of Agriculture
Umudike, Umuahia, Abia State, Nigeria

*Address all correspondence to: obianyo.jude@mouau.edu.ng

IntechOpen

References

[1] Peters K. Polymer optical fiber sensors - A review. Smart Materials and Structures 2011, 20(1), 013002. IOP Publishing Ltd.

[2] Alwis L, Sun T, Grattan KTV. Developments in optical fiber sensors for industrial applications. Optics and Laser Technology 2016, vol. 78, Part A, pp. 62-66. doi: 10.1016/j.optlastec.2015.09.004.

[3] Mishra V, Singh N, Tiwari U, Kapur P. Fiber grating sensors in medicine: Current and emerging applications. Sensors and Actuators A: Physical 2011, 167(2): 279-290. doi: 10.1016/j.sna.2011.02.045

[4] Theodosiou A, Komodromos M, Kalli K. Carbon cantilever beam health inspection using a polymer fiber Bragg grating array. Journal of Lightwave Technology 2018, 36(4): 986-992.

[5] Leal-Junior AG, Frizera A, Jose Pontes M. Sensitive zone parameters and curvature radius evaluation for polymer optical fiber curvature sensors. Optics and Laser Technology 2018, 100 (1): 272-281. doi: 10.1016/j.optlastec.2017.10.006.

[6] Zhong N, Liao Q, Zhu X, Zhao M, Huang Y, Chen R. Temperature-independent polymer optical fiber evanescent wave sensor. Scientific Reports 2015, 5: 11508. doi: 10.1038/srep11508.

[7] Leal-Junior A, Frizera-Neto A, Marques C, Pontes MA. A polymer optical fiber temperature sensor based on material features. Sensors 2018, 18, 301; https://doi.org/10.3390/s18010301.

[8] Stefani A, Andresen S, Yuan W, Herhold-Rasmussen, N, Bang O. High sensitivity polymer optical fiber-bragg-grating-based accelerometer. IEEE Photonics Technology Letters 2012, 24(9): 763-765. doi: 10.1109/LPT.2012.2188024.

[9] Vilarinho D, Theodosiou A, Leitao C, Leal-Junior AG, Domingues M, Kalli K, Andre P, Antunes P, Marques C. POFBG-embedded cork insole for plantar pressur monitoring. Sensors 2017, 17(12): 2924. doi: 10.3390/s17122924.

[10] Chen Z, Lau D, Teo JT, Ng SH, Yang X, Kei PL. Simultaneous measurement of breathing rate and heart rate using a microbend multimode fiber optic sensor. Journal of Biomedical Optics 2014, 19(5): 057001. doi: 10.1117/1.JBO.19.5.057001. PMID: 24788372.

[11] Kronenberg P, Rastogi PK, Giaccari P, Limberger PK. Relative humidity sensor with optical fiber Bragg gratings. Optics Letters 2002, 27(16): 1385-1387.

[12] Grandal T, Zornoza A, Lopez A, Fraga S, Sun T, Grattan KTV. Analysis of fiber optic sensor embedding in metals by automatic and manual TIG welding. IEEE Sensors Journal 2019, 19 (17), pp. 7425-7433. doi: 10.1109/JSEN.2019.2916639.

[13] Grandal T, Zornoza A, Fraga S, Castro G, Sun T, Grattan KTV. Laser cladding-based metallic embedding technique for fiber optic sensors. Journal of Lightwave Technology 2018, 36(4): 1018-1025. doi: 10.1109/JLT.2017.2748962.

[14] Pan M, Sheng J, Liu J, Shi Z, Jiu L. Design and verification of humidity sensors based on magnesium oxide micro-arc oxidation film layer. Sensors 2020, 20, 1736-1748. doi: 10.3390/s20061736.

[15] Shaikh SF, Ghule BG, Shinde VP, Raut SD, Gore SK, Ubaidullah M, Mane RS, Al-Enizi AM. Continuous

hydrothermal flow-inspired synthesis and ultra-fast ammonia and humidity room-temperature sensor activities on WO_3 nanobricks. Materials Research Express 2020, 7, 015076. doi: 10.1088/2053-1591/ab67fc.

[16] Jyoti GDV. Enhanced room temperature sensitivity of Ag-CuO nanobricks/reduced grapheme oxide composite for NO_2. Journal of Alloys and Compounds 2019, 806: 1469-1480. doi: 10.1016/j.jallcom.2019.07.355.

[17] Ijima S. Helical microtubules of graphitic carbon. Nature 1991, 354: 56-58.

[18] Baughman RH, Zakhidov AA, De Heer WA. Carbon nanotubes - The route toward applications. Science 2002 Aug 2; 297(5582): 787-792. doi: 10.1126/science.1060928. PMID: 12161643.

[19] Schnorr JM, Swager TM. Emerging applications of carbon nanotubes. Chemistry of Materials 2011, 23(3): 646-657. doi: 10.1021/cm102406h.

[20] Arunachalam S, Gupta AA, Izquierdo R, Nabki F. Suspended carbon nanotubes for humidity sensing. Sensors 2018, 18, 1655: doi: 10.3390/s18051655.

[21] Singh R, Yadav A, Gautam C. Synthesis and humidity sensing investigations of nanostructured $ZnSnO_3$. Journal of Sensor Technology 2011, 1(4). Doi: 10.4236/jst.2011.14016.

[22] Zak AK, Shirmahd H, Mohammadi S, Banihashemian SM. Solvothermal synthesis of porous Fe3O4 nanoparticles for humidity sensor application. Materials Research Express 7 (2020) 025001.

[23] Zafar Q, Azmer MI, Al-Sehemi AG, Al-Assiri MS, Kalam A, Sulaiman K. Evaluation of humidity sensing properties of TMBHPET thin film

embedded with spinel cobalt ferrite nanoparticles. Journal of Nanoparticle Research 2016, 18(7). doi: 10.1007/s11051-016-3488-9.

[24] Azmer MI, Aziz F, Ahmad Z, Raza E, Najeeb MA, Fatima N, Bawazeer TM, Alsoufi MS, Shakoor R, Sulaiman K. Compositional engineering of VOPcPhO-TiO2 nano-composite to reduce absolute threshold value of humidity sensors. Talanta 2017, 174: 279-284. ISSN 0039-9140.

[25] Najeeb MA, Ahmad Z, Shakoor RA. Organic thin-film capacitive and resistive humidity sensors: A focus review. Advanced Materials Interfaces 2018, 1800969. doi: 10.1002/adm.201800969.

[26] Harrey PM, Ramsey BJ, Evans PSA, Harrison DJ. Capacitive-type humidity sensors fabricated using the offset lithographic printing process. Sensors and Actuators B: Chemical 2002, 87(2): 226-232.

[27] Ralston ARK, Klein CF, Thoma PE, Denton DD. A model for the relative environmental stability of a series of polyimide capacitive humidity sensors. Sensors and Actuators B: Chemical 1996, 34(1-3): 343-348.

[28] Smit H, Kivi R, Vomel H, Paukkunen A, In monitoring atmospheric water vapour (Ed: N. Kampfe), Springer, New York, NY 2013, pp. 11-38.

[29] Sakai Y, Sadaoka, Y. Matsuguchi M. Humidity sensors based on polymer thin films. Sensors and Actuators B 1996, 35 (1-3): 85-90. doi: 10.1016/S0925-4005 (96)02019-9.

[30] Kelleners TJ, Soppe RWO, Robinson DA, Schaap MG, Ayars JE, Skaggs TH. Calibration of capacitance probe sensors using electric circuit theory. Soil Science Society of America Journal 2004, 68: 430-439.

[31] Placidi P, Gasperini L, Grassi A, Cecconi M, Scorzoni A. Characterization of low-cost capacitive soil moisture sensors for IoT networks. Sensors 2020, 20: 3585. doi: 10.3390/s20123585.

[32] Butler SZ, Hollen SM, Cao L, Cui Y, Gupta JA, Gutierrez HR, Heinz TF, Hong SS, Huang J, Ismach AF, Johnston-Halperin E, Kuno M, Plashnista VV, Robinson RD, Ruoff RS, Salahuddin S, Shan J, Shi L, Spencer MG, Terrones M, Windl W, Goldberger JE. Progress, challenges, and opportunities in two-dimensional materials beyond graphene. ACS Nano 2013, 7(4): 2898-2926.

[33] Cho B, Hahm MG, Choi M, Yoon J, Kim AR, Lee Y-J, Park S-G, Kwon J-D, Kim CS, Song M, Jeong Y, Nam K-S, Lee S, Yoo TJ, Kang CG, Lee BH, Ko HC, Ajayan PM, Kim D-H. Charge-transfer-based gas sensing using atomic-layer MoS$_2$. Scientific Report 5, Article number 8052 (2015)

[34] Awais M, Khan MU, Hassan A, Bae J, Chatta TE. Printable highly stable and superfast humidity sensor based on two dimensional molybdenum diselenide. Scientific Reports 2020, 10-5509. Doi: 10.1038/s41598-020-62397.

[35] Chou KS, Lee CH, Liu BT. Effect of microstructure of ZnO nanorod film on humidity sensing. J. Am. Ceram. Soc. 2016, 99, 531-535. doi: 10.1111/jace.13994.

[36] Duan Z, Xu M, Li T, Zhang Y, Zou H. Super-fast response humidity sensor based on La0. 7Sr0. 3MnO3 nanocrystals prepared PVP-assisted sol-gel method. Sensors and Actuators B Chemical 2018, 258, 527-534. doi: 10.1016/j.snb.2017.11.169.

[37] Fei T, Dai J, Jiang K, Zhao H, Zhang T. Stable cross-linked amphiphilic polymers from a one-pot reaction for application in humidity sensors. Sensors and Actuators B Chemical 2016, 227, 649-654. doi: 10.1016/j.snb.2016.01.038.

[38] Agarwal S, Sharma G. Humidity sensing properties of of (Ba, Sr) TiO$_3$ thin films grown by hydrothermal-electrochemical method. Sensors and Actuators B Chemical 2002, 85, 205-211. doi: 10.1016/S0925-4005(02000109-0.

[39] Wang R, Wang D, Zhang Y, Zheng X. Humidity sensing properties of Bi$_{0.5}$(Na$_{0.85}$K$_{0.15}$)$_{0.5}$Ti$_{0.97}$Zr$_{0.03}$O$_3$ microspheres: Effect of A and B sites co-substitution. Sensors and Actuators B Chemical 2014, 190, 305-310. doi: 10.1016/j.snb.2013.08.048.

[40] Ghatak S, Pal AN, Ghosh A. Nature of electronic states in atomically thin MoS$_2$ field-effect transistors. ACS Nano 2011, 5(10): 7707-7712. doi: 10.1021/nn202852

[41] Rout CS, Joshi PD, Kashid RV, Joag DS, More MA, Simbek AJ, Washington M, Nayak SK Late DJ. Superior field emission properties of layered WS2-RGO nanocomposites. Scientific Reports 3, 2013, Art. No. 3282. doi: 10.1038/srep03282.

[42] Braga D, Lezama IG, Berger H, Morpurgo AF. Quantitative determination of the band gap with ambipolar ionic liquid-gated transistors. Nano letters 2012, 12(10): 5218-5223. doi: 10.1021/nl302389d.

[43] Georgiou T, Jalil R, Belle BD, Britnell L, Gorbachev RV, Morozov SV, Kim Y-J, Gholinia A, Haigh SJ, Makarovsky O, Eaves L, Ponomarenko LA, Geim AK, Novoselov KS, Mishchenko A. Vertical field-effect transistor based on graphene-WS2 heterostructures for flexible and transparent electronics. Nature Nanotechnology 2013, 8, 100-103. doi: 10.1038/nnano.2012.224.

[44] Late DJ, Huang Y-K, Liu B, Acharya J, Shirodkar SN, Luo J, Yan A,

Charles D, Waghmare UV, Dravid VP, Rao CNR. Sensing behavior of atomically thin-layered MoS$_2$ transistors. ACS Nano 2013, 7(6): 4879-4891. doi: 10.1021/nn400026u.

[45] Late DJ, Liu B, Ramakrishna Matte HSS, Dravid VP, Rao CNR. Hysteresis in single-layer MoS$_2$ field effect transistors. ACS Nano 2012, 6(6): 5635-5641. doi:10.1021/nn301572c.

[46] Late DJ, Liu B, Ramakrishna Matte HSS, Rao CNR, Dravid PV. Rapid characterization of ultrathin layers of chalcogenides on SiO$_2$/Si substrates. Advanced Functional Materials 2012, 22 (9): 1894-1905. doi: 10.1002/adfm.201102913.

[47] Larentis S, Fallahazad B, Tutuc E. Field effect transistors and intrinsic mobility in ultra-thin MoSe$_2$ layers. Applied Physics Letters 2012, 101, 223104. doi: 10.1063/1.4768218.

[48] Bernadi M, Palummo M, Grossmann JC. Extraordinary sunlight absorption and one nanometer thick photovoltaics using two-dimensional monolayer materials. Nano Letters 2013, 13(8): 3664-3670. doi: 10.1021/nl401544y.

[49] Eftekhari A. Molybdenum diselenide (MoSe$_2$) for energy storage, catalysis, and optoelectronics. Applied Materials Today 2017, 8: 1-17. doi: 10.1016/j.apmt.2017.01.006.

[50] Kong D, Wang H, Cha JJ, Pasta M, Koski KJ, Yao J, Cui Y. Synthesis of MoS$_2$ and MoSe$_2$ films with vertically aligned layers. Nano Letters 2013, 13(3): 1341-1347. doi: 10.1021/nl400258t.

[51] Kioseoglu G, Hanbicki AT, Currie M, Friedman AL, Jonker BT. Optimal polarization and intervalley scattering in single layers of MoS$_2$ and MoSe$_2$. Scientific Reports 6, Article number 25041 (2016).

[52] Chang Y-H, Zhang W, Zhu Y, Han Y, Pu J, Chang J-K, Hsu W-T, Huang J-K, Hsu C-L, Chiu M-H, Takenobu T, Li H, Wu C-I, Chang W-H, Wee ATS, Li L-J. Monolayer MoSe2 grown by chemical vapor deposition for fast photodetection. ACS Nano 2014, 8 (8): 8582-8590. doi: 10.1021/nn503287m.

[53] Chappanda KN, Shekhah O, Yassine O, Patole SP, Eddaoudi M, Salama KN. The quest for highly sensitive QCM humidity sensors: The coating of CNT/MOF composite sensing films as case study Sensors and Actuators B 2018, 257: 609-619. doi: 10.1016/j.snb.2017.10.189

[54] Farahani H, Wagiran R, Hamidon MN. Humidity sensors principle, mechanism, and fabrication technologies: A comprehensive review. Sensors 2014, 14, 7881. doi: 10.3390/s140507881.

[55] Pal BN, Chakravorty D. Humidity sensing by composites of glass ceramics containing silver nanoparticles and their conduction mechanism. Sensors and Actuators B Chemical 2006, 114, 1043-1051.

[56] Jeseentharami V, Reginamary L, Jeyaraj B, Dayalan A, Nagaraja KS. Nanocrystalline spinel Ni$_x$Cu$_{0.8-x}$Zn$_{0.2}$Fe$_2$O$_4$: A novel material for humidity sensing. Journal of Material Science 2011, 47: 3529-3534.

[57] Mukode S, Futata H. Semiconductive humidity sensor. Sensors and Actuators 1989, 16: 1-11.

[58] Shimizu Y. Humidity-sensitive characteristics of La^{3+}-doped and undoped SrSnO$_3$. Journal of the Electrochemical Society 1989, 136: 1206.

[59] Jain M, Bhatnagar M, Sharma G. Effect of Li$^+$ doping on ZrO$_2$-TiO$_2$ humidity sensor. Sensors and Actuators B Chemical 1999, 55: 180-185.

[60] Hu X, Gong J, Zhang L, Yu JC. Continuous size tuning of monodisperse ZnO colloidal nanocrystal clusters by a microwave-polyol process and their application for humidity sensing. Advanced Materials 2008, 20: 4845-4850.

[61] Rezlescu N, Doroftei C, Popa PD. Humidity-sensitive electrical resistivity of $MgFe_2O_4$ and $Mg_{0.9}Sn_{0.1}Fe_2O_4$ porous ceramics. Romanian Journal of Physics 2007, 52: 353-360.

[62] Qiu Y, Yang S. ZnO nanotetrapods: controlled vapour-phase synthesis and application for humidity sensing. Advanced Functional Materials 2007, 17: 1345-1352.

[63] Shimizu Y, Okada H, Arai H. Humidity-sensitive characteristics of porous La-Ti-V-O glass-ceramics. J. Am. Ceram. Soc. 1989, 72: 436-440.

[64] Smith AD, Li Q, Anderson A, Vyas A, Kuzmenko V, Haque M, Staaf LGH, Lundgren R, Enoksson P. Toward CMOS compatible wafer-scale fabrication of carbon-based microsupercapacitors for IoT. Journal of Physics 2018: Conference Series, volume 1052, 012143, 17th International Conference on Micro and Nanotechnology for Power Generation and Energy Conversion Applications (PowerMEMS 2017) 14-17 November 2017, Kanazawa, Japan.

[65] Hester JGD, Kimionis J, Tentzeris MM. Printed motes for IoT wireless networks: State of the art, challenges and outlooks. IEEE Transactions on Microwave Theory and Techniques 2017, 65(5): 1819.

[66] Balpande SS, Pande RS, Patrikar RM. Design and low cost fabrication of green vibration energy harvester. Sensors and Actuators A: Physical 2016, 251: 134-141.

[67] Rahman MM, Khan SB, Asiri AM. Fabrication of smart chemical sensors based on transition-doped-semiconductor nanostructure materials with μ-chips. PLoS One, 9(1): e85036. doi: 10.1371/journal.pone.0085036.

[68] Korotcenkov G. The role of morphology and crystallographic structure of metal oxides in response of conductometric-type gas sensors. Materials Science and Engineering R Reports 2008, 61(1): 1-39. (2008). doi: 10.1016/j.mser.2008.02.001

[69] Munneb-ur-Rahman, Shah G, Ullah, A. Resistive- and capacitive-type humidity and temperature sensors based on a novel caged nickel sulfide for environmental monitoring. J. Mater. Sci: Mater Electron 31, 3557-3563. Doi: 10.1007/s10854-020-02904-y.

[70] Joshi SR, Kim B, Kim S-K, Song W, Park K, Kim G-H, Shin H. Low-cost and fast-response resistive humidity sensor comprising biopolymer-derived carbon thin film and carbon microelectrodes. Journal of the Electrochemical Society 2020, 167: 147511.

[71] Choi S-J, Yu H, Jang J-S, Kim M-H, Kim S-J, Jeong HS, Kim I-D. Nitrogen-doped single graphene fiber with platinum water dissociation catalyst for wearable humidity sensor. Small 2018, Vol. 14 Issue 1/e1703934. doi: 10.1002/smll.201703934.

[72] Barnes J, Stephenson R, Welland M, Gerber C, Gimzewski J. Photothermal spectroscopy with femtojoule sensitivity using a micromechanical device. Nature 1994, 372(6501): 79-81.

[73] Yoshikawa G, Akaiyama T, Gautsch S, Vettiger P, Rohrer H. Nanomechanical membrane-type surface stress sensor, Nano Letters 2011, 11(3): 1044-1048.

[74] Li M, Tang HX, Roukes ML. Ultra-sensitive NEMS-based cantilevers for sensing, scanned probe and very high-frequency applications. Nature Nanotechnology 2007, 2(2): 114-120.

[75] Takahashi T, Hizawa T, Misawa N, Taki M, Sawada K, Takahashi K. Surface stress sensor based on MEMS Fabry-Perot interferometer with high wavelength selectivity for label-free biosensing. Journal of Micromechanics and Microengineering 2018, 28(5): Art No. 054002.

[76] Seena V, Fernandes A, Pant P, Mukherji S, Rao VR. Polymer nanocomposite nanomechanical cantilever sensors: Material characterization, device development and application in explosive vapour detection. Nanotechnology 2011, 22 (29), Art No. 295501.

[77] Huang X, Manolidis, M, Jun SC, Hone J. Nanomechanical hydrogen sensing. Applied Physics Letters 2005, 86(14), Art No. 143104.

[78] Pei J, Tian F, Thundat T. Glucose biosensor based on the microcantilever. Analytical Chemistry 2004, 76(2): 292-297.

[79] Hossain MM, Toda M, Hokama T, Yamazaki M, Moorthi K, Ono T. Piezoresistive nanomechanical humidity sensors using internal stress in-plane of Si-polymer composite membranes. IEEE Sensors Letters 2019, 3(2), 2500404.

[80] Staruch M, Kassner C, Fackler S, Takeuchi I, Bussmann K, Lofland SE, Dolabdjian C, Lacomb R, Finkel P. Effects of magnetic field and pressure in magnetoelastic stress reconfigurable thin film resonators. Applied Physics Letters 2015, 107, 032909-4.

[81] Grimes CA, Mungle CS, Zeng K, Jain MK, Dreschel WR, Paulose M, Ong KG. Wireless magnetoelastic resonance sensors: A critical review. Sensors 2002, 2: 294-313.

[82] Stoyanov PG, Grimes CA. A remote query magnetoelastic viscosity sensor. Sensors and Actuators A 2000, 80: 8-14.

[83] Grimes CA, Sommath CR, Rani S, Qingyum C. Theory, instrumentation and applications of magnetoelastic resonance sensors: A review. Sensors 2011, 11: 2809-2844.

[84] Atalay S, Izgi T, Kolat VS, Erdemoglu S, Inan OO. Magnetoelastic humidity sensors with TiO2 nanotube sensing layers. Sensors 2020, 20, 425. doi: 10.3390/s20020425.

[85] Pang PF, Zhang YL, Ge ST, Cai QY, Yao SZ, Grimes CA. Determination of glucose using bienzyme layered assembly magnetoelastic sensing device. Sensors and Actuators B Chemical 2009, 136: 310-314.

[86] Kouzoudis D, Grimes CA. The frequency response of magnetoelastic sensors to stress and atmospheric pressure. Smart Materials and Structures 2000, 71: 3822-3824.

[87] Atalay S, Kolat VS, Bayri N, Izgi T. Magnetoelastic sensor studies on amorphous magnetic FeSiB wire and the application in viscosity measurement. Journal of Superconductivity and Novel Magnetism 2016, 29: 1551-1556.

[88] Jain MK, Cai YQ, Grimes CA. A wireless micro-sensor for simultaneous measurement of pH, temperature and pressure. Smart Materials and Structures 2001, 10:347-353

[89] Cai YQ, Cammers-Goodwin A, Grimes CA. A wireless remote query magnetoelastic CO_2 sensor. Journal of Environmental Monitoring 2000, 2: 556-560

[90] Grimes CA, Kouzoudis D, Dickey EC, Qian D, Anderson MA, Shahidian R, Lindsey M, Green L. Magnetoelastic sensors in combination with nanometer-scale honey combed thin film ceramic TiO_2 for remote query measurement of humidity. Journal of Applied Physics 2000, 87: 5341-5343.

[91] Jain MK, Schmidt S, Ong KG, Mungle C, Grimes CA. Magnetoacoustic remote query temperature and humidity sensors. Smart Materials and Structures 2000, 9: 502-510.

[92] Grimes CA, Kouzoudis D. Remote query measurement of pressure, fluid-flow velocity and humidity using magnetoelastic thick-film sensors. Sensors and Actuators A 2000, 84: 205-212.

[93] Zhang K, Zhang L, Yuesheng C. Mass load distribution dependence of mass sensitivity of magnetoelastic sensors under different resonance modes. Sensors 2015, 15: 20267-20278.

[94] Saiz PG, Gandia D, Lasheras A, Sagasti A, Quintana I, Fdez-Gubieda ML, Gutierrez J, Arriortua MI, Lopes AC. Enhanced mass sensitivity in novel magnetoelastic resonators geometrics for advanced detection systems. Sensors and actuators B Chemical 2019, 296, 126612.

[95] Yao C, Zhu T, Tang J, Wu R, Chen Q, Chen M, Zhang B, Huang J, Fu W. Hybridization assay of hepatitis B virus by QCM peptide nucleic aci biosensor. Biosensors and Bioelectronics 2008, 23(16): 897-885. doi: 10.1016/j.bios.2007.09.003.

[96] Sheng Z, Huang M, Xiao C, Zhang Y, Zeng X, Wang PG. Non-labeled quartz crystal microbalance biosensor for bacterial detection using carbohydrate and lectin recognitions. Analytical Chemistry 2007, 79(6): 2312-2319. Doi: 10.1021/ac0619861.

Chapter 4

MEMS Humidity Sensors

Ahmad Alfaifi, Adnan Zaman and Abdulrahman Alsolami

Abstract

This chapter reviews MEMS humidity sensors fabricated using microfabrication technologies. It discusses the operation principle, different designs, and the fabrication technologies for the different sensing mechanisms. Sensing humidity using capacitive sensors is first reviewed with a highlight on the different sensing materials and how their permittivity and physical parameters affect the sensor performance. Then the chapter discusses the piezoelectric humidity sensing method, wherein piezoelectric sensors the dynamic mode measurement is used. In these sensors, the mass changes corresponding to the humidity, resulting in resonance frequency shift and amplitude change. Finally, the chapter reviews the resistive humidity sensors where the change in the resistivity of various materials is used as an indication of humidity change in the environment.

Keywords: Humidity sensors, MEMS, capacitive sensors, piezoelectric sensors, resistive sensors

1. Introduction

Miniaturizing devices and systems are the key factors of today's technology advancement. Humidity sensor developments in all its categories are tightly related to micromachining technology development. Such developments allow the advantages of high performance, low power consumption, cost reduction, and the capability to fabricate batches of devices and systems that are needed for today's technology [1].

Micro Electro Mechanical Systems (MEMS) are miniature devices fabricated using micro fabrication processes that are used for both sensing and actuation. When used in sensing, their mechanical properties are employed to generate electrical signals that indicate the measured parameter. When used in actuation, the electrical signal is used to produce micro movements. MEMS have been widely used in different sensing applications ranging from environmental sensing, e.g., temperature, to motion and position and detection, e.g., accelerometers [2].

Humidity, in general, can be measured using absolute humidity, dew point, mixing ratio, and relative humidity. Among these measurements, relative humidity is the most common measure used in the literature [3–7]. In a closed system, relative humidity (RH) is defined as the total vapor pressure in volume to the pressure where the water vapor saturated at a given temperature, and its formula is generally expressed as:

$$\%RH = \frac{P_t}{P_s} \tag{1}$$

where P_t is the total pressure and P_s is the saturated pressure [3].

This chapter will review the three main sensing schemes usually used for humidity sensing in MEMS devices, starting with capacitive sensing with its sensing principle and fabrication process. The piezoelectric sensing method is then detailed, followed by a description of the resistive sensing. In each part, the sensing mechanism will be briefly reviewed with a glance at the fabrication method. The chapter is then concluded by a comparison between the three sensing schemes.

2. Capacitive humidity sensing technology

Figure 1a depicts the basic design of a capacitive humidity sensor. This sensing method depends mainly on changing the permittivity of a sensing material due to its absorbance of water vapor molecules leading to a change in the capacitance value. As the sensing materials are exposed to the environment, water vapor molecules enter their pores resulting in a change of the sensing materials' permittivity. This change occurs because the water molecules have high permittivity due to their polar structure compared to the permittivity of the sensing materials. The sensitivity of the used layer can be attributed to several parameters, e.g., pores sizes, layer thickness, and area exposed to the environment [8]. Capacitive sensing has several advantages over other sensing methods. It has a simple readout circuit, low power consumption, and nonmoving structure [9].

To calibrate the capacitance of the sensing element, another capacitive structure is fabricated next to it as a reference, **Figure 1b**. The sensing materials in this reference capacitor are completely covered by the top electrode and are not exposed to the environment. Hence, its permittivity can be referenced as it does not absorb any water vapor. The performance of the humidity sensor can be improved by adding a heating element beneath it. The heating element increases the sensing materials' temperature, which increases the humidity diffusion constant and hence the response time. It can also be used to reset the sensor faster to the dry state [9]. Notice that heating the sensor will affect its dielectric constant, so compensation must be made to the readings to account.

2.1 Design and operation

Humidity can be sensed using a capacitive scheme by using two parallel electrodes with a material sensitive to water vapor as the insulating layer between the two electrodes, **Figure 2a**. The device capacitance is given by [8]:

$$C = \varepsilon_0 \varepsilon_r \frac{A}{d} \tag{2}$$

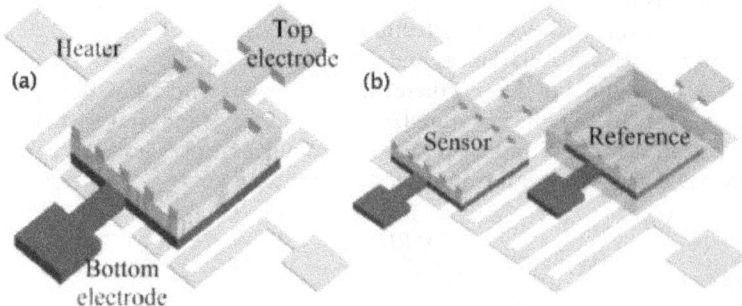

Figure 1.
A 3D model of a capacitive humidity sensor as a (a) standalone, and (b) next to a capacitive reference [8].

where ε_0 is the air permittivity, ε_r is the relative permittivity of the sensing material, A is the capacitive area, and d is the capacitive gap, i.e., sensing material thickness.

It is important that the device structure exposes the sensing material from its side surfaces in order to allow the sensing material between the electrodes to absorb moisture effectively, **Figure 2b**. If the sensing material is left without patterning, most of the moisture will be absorbed in the part that is not sandwiched between the electrodes, leading to low sensitivity [9].

There are several parameters that are used to characterize the capacitive sensor performance, e.g., response/recovery time, range, sensitivity, etc. The sensitivity of these sensors is usually linear and largely determined by device design structure design, and it can be expressed as [9]:

$$S = \frac{\Delta C/C}{\Delta RH} \tag{3}$$

where ΔC is the capacitance change due to humidity, C is the nominal capacitance, and ΔRH is the relative humidity variation.

For a quick response time and better sensitivity, the sensing material thickness should be minimized. Furthermore, the thickness of the electrodes should be minimized for better sensitivity and to reduce the parasitic capacitance, but that means additional fabrication steps that are needed to form the bonding pads [10–12]. Capacitive sensors can measure the humidity over the entire range from 0–100%, which makes them preferable in most applications [9].

2.2 Fabrication

Simple capacitive humidity sensors can be easily fabricated in a 3-mask process [9], **Figure 3**. The process starts with a handle silicon wafer covered with a thin layer of silicon dioxide to create an insulation layer that prevents short circuiting the capacitive electrodes when they are deposited later. Then a metallic layer, e.g., aluminum, is deposited on the wafer to form the bottom electrode. If the same layer will be used to create the pad for the wire bonding then its thickness should be around 300 nm to be able to stand the bonding step without punching through the layer and losing the signal connection. This metallic layer is then patterned using the first photolithography mask. Next, the sensing materials were deposited and patterned using the second photolithography mask. If the sensing materials are a polymer, e.g., polyimide, it can be spun and then cured before patterning it using a hard mask and an oxygen reactive ion etching step. The thickness of the sensing materials determines the performance of the device, which makes it critical to be

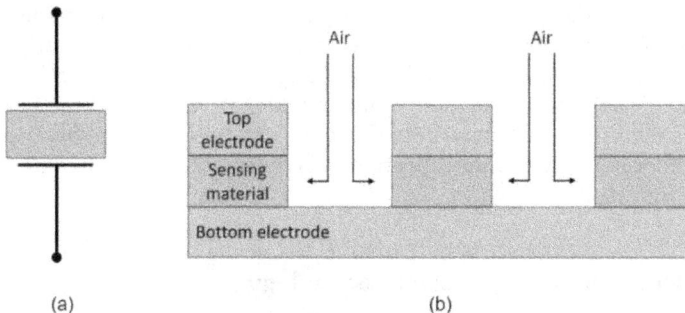

(a)   (b)

Figure 2.
(a) Capacitive humidity sensor sensing scheme; and (b) a cross-section of the fabricated sensor [9].

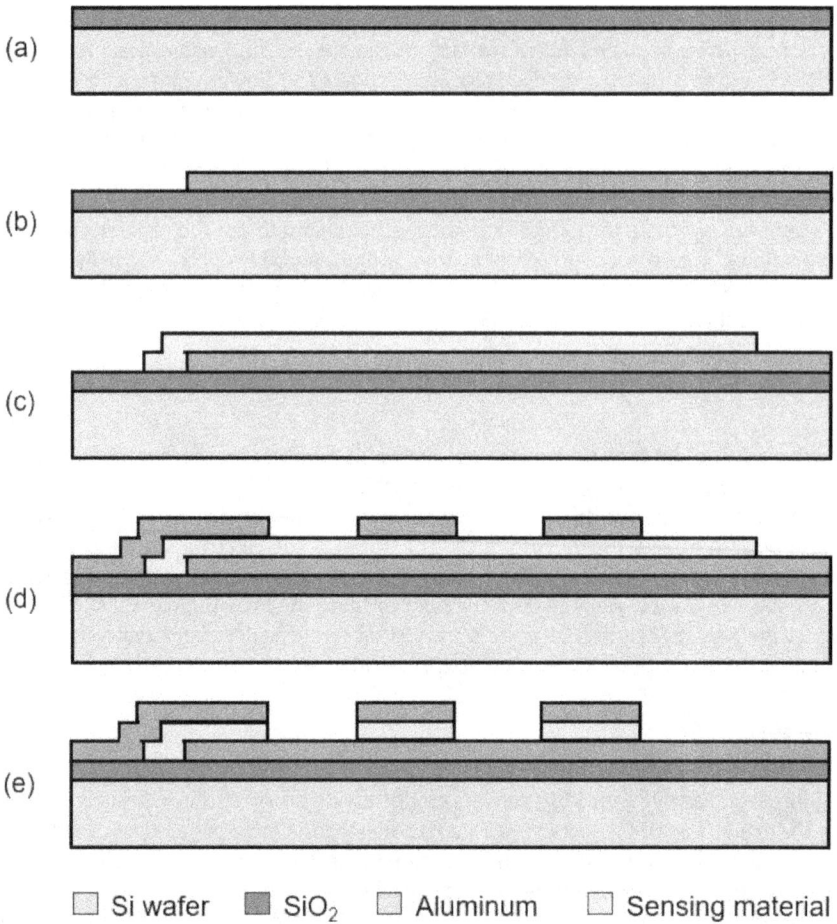

Figure 3.
Cross-section of a simple capacitive sensing fabrication process flow [9], depicting the silicon wafer after (a) being covered with silicon oxide, (b) depositing and patterning the bottom electrode, (c) spinning and patterning the sensing material, (d) depositing and patterning the top electrode, and (e) etching the sensing materials to create the final structure.

controlled accurately, especially for the mass production of these sensors. The top electrode is then deposited and patterned using the third mask. If the same layer is to be used for the top electrode wire bonding, then the layer should be around 300 nm thick as well and the material must be a metal that can be wire bonded to, e.g., aluminum. In the last step, the sensing material is then etched in anisotropic etching step to expose the sensing surfaces and form the final structure.

There are a variety of materials that can be used to sense humidity; however, polymers and ceramics are the most common materials in MEMS humidity sensors [8]. Porous semiconductors have been demonstrated to sense humidity. However, they have poor linearity and operate efficiently only in the range of relative humidity higher than 10–20%.

3. Piezoelectric humidity sensing technology

RF MEMS is one of the MEMS categories used to transmit radio frequency signals that operate in ultra-high frequency. A micromechanical resonator can be used as a sensor to couple the energy in and out the resonator's mechanical structure by actuating and sensing motion out of the mechanical resonator. The piezoelectric transduction mechanism is used to transfer the energy between the mechanical/electrical domains in the micromechanical system. Piezoelectric materials like ZnO can generate an electrical charge in response to applied external mechanical stress (force). This happens because the internal reticular electric polarization from piezoelectric materials is perturbed by mechanical means, and an electrical response can be generated because of the induced dielectric displacement. This behavior is called the direct piezoelectric effect [13].

On the contrary, the converse piezoelectric effect appears after applying an external electric field to a material that generates a mechanical deformation across the piezoelectric materials, which is directly proportional to both the electric field's strength and the equivalent acoustic velocity (V_{eq}) within the bulk piezoelectric material layer. The piezoelectric layer's internal electric polarization is affected by the mechanical deformation of the resonance frequency, and the piezoelectric effect can be calculated from the output current i_0. The piezoelectric effect can be closely governed by the following Equations [14]:

$$T = e^S \cdot S - p \cdot E \tag{4}$$

$$D = p \cdot S + \varepsilon^S \cdot E \tag{5}$$

where S, T, D, and E represent the strain, stress, electric displacement, and electric field, respectively. Also, e^S is the elastic stiffness at a constant electric field, ε^S is the permittivity at a constant strain, and p is the piezoelectric constant [14].

MEMS sensors based on metal oxide semiconductors such as zinc oxide (ZnO) and aluminum oxide (Al_2O_3) have become one of the most attractive sensors for gas detecting applications [14]. Thin-film piezoelectric-on-substrate (TPoS) resonators based on ZnO have a considerable prospect to be used in mass or gas sensing applications. TPoS resonators are strategically coupled with low loss substrates like Si to have higher acoustic velocities and to store more energy which results in increasing the equivalent acoustic velocity to be higher than typical piezoelectric resonators [15]. Based on a previous study [16], the effects of mass loading on TPoS based resonators have shown high sensitivity of the first and fourth-order contour mode ZnO-on-Si MEMS resonator-based mass sensor, which provides massive potential in mass sensing applications.

3.1 Piezoelectric MEMS mass resonator

Piezoelectric MEMS mass resonator consists of top and bottom electrodes sandwiching a structural layer (resonator body) that can be silicon, nickel, diamond, or any other low acoustic loss structural material. The equivalent acoustic velocity can be found by [14]:

$$V_{eq} = \sqrt{\frac{E_1 T_1 + E_2 T_2 + \dots + E_m T_m}{(\rho_1 T_1 + \rho_2 T_2 + \dots + \rho_m T_m)(1 - \sigma^2)}} \tag{6}$$

where m is the number of the stacked layers; T is the thickness of each material; ρ, E and σ indicate the density, Young's Modulus, and Poisson's ratio of the stacked piezoelectric resonator structural materials [14]. The piezoelectric sensor is designed to be operated by applying a mechanical force to the body of the sensor which will excite the resonator system into the designed frequency mode. MEMS resonator sensor is extremely sensitive and can probably provide the most accurate measurement comparing with other technology. The piezoelectric MEMS resonator's output is a frequency that can be easily measured and monitored with very high accuracy using a vector network analyzer (VNA), as shown in **Figure 4a**. The frequency of the device all depends on the geometrical parameters as well as the material properties of the structure. The piezoelectric resonator can be used in gas sensing and humidity sensor applications. This resonator is sensitive enough to detect and sense any slight mass changes [17].

3.1.1 Design and operation

There are two modes of operation for MEMS sensors: static and dynamic modes. In the static mode, the resonator remains stationary where its deflection and actuation depend only on the variation of the surface stress. In the dynamic mode, the changes occur on the resonator's mechanical stiffness, and the mass variation results in resonance frequency shift and amplitude change [18]. Dynamic mode is the most commonly used technique for gas and humidity sensing applications.

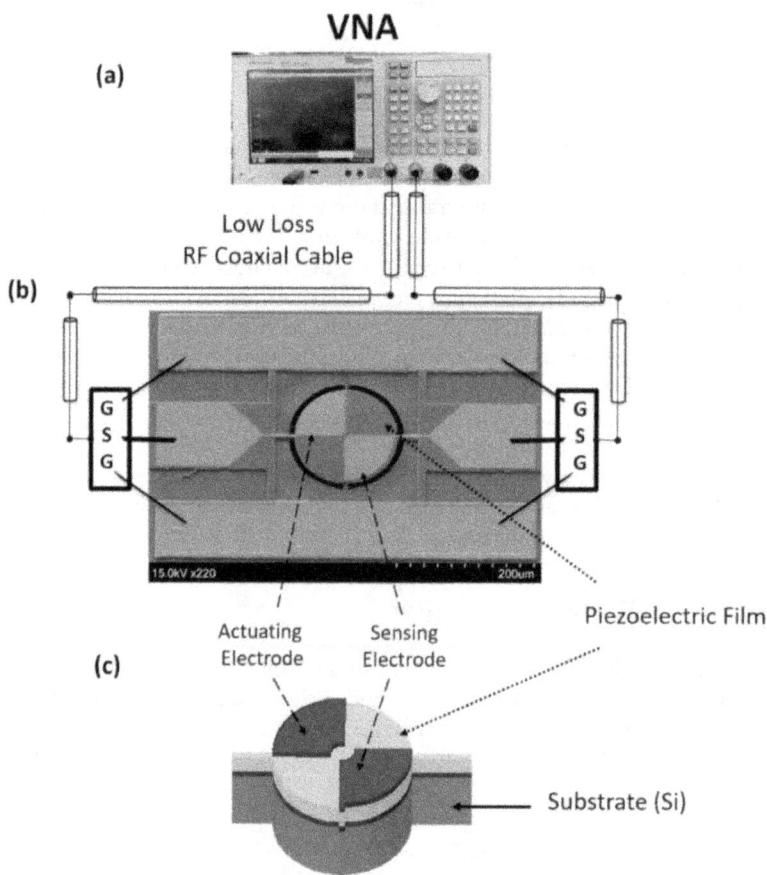

Figure 4.
(a) Illustration of RF measurement set-up for piezoelectric mass sensor device to measure the frequency response; (b) SEM image showing 2D illustration of the dynamic mass of a 1 st contour mode disk resonator for piezoelectric mass sensor device; (c) 3D illustration of sensing film, actuating and sensing electrodes for piezoelectric mass sensor device [14].

For humidity sensing application, MEMS resonator can be coated with a water or gas absorbing layer such as Metal–Organic Frameworks (MOFs). As the humidity increases, the value of the resonance frequency of the piezoelectric sensor will be shifted down due to the increase of the mass. The mass-loading effect can be calculated by monitoring the resonance frequency change of MEMS resonator, which is proportional to the ratio between the mode frequency and the equivalent mass given by [19]:

$$\Delta f = \frac{f_n}{2m_e} \Delta m \tag{7}$$

where m_e is the effective mass of the resonator (also known as the equivalent dynamic mass of the resonator), Δf is the measurable resonance frequency shift due to the loaded mass, Δm is the change of mass, and f_n is the resonance frequency of the n-order contour mode [19].

The piezoelectric MEMS resonator sensor's working principle depends on the change in the frequency of the resonator structure due to the mass loading effect. Air can only absorb a certain amount of water vapor. This amount highly depends on the temperature. Piezoelectric MEMS sensor measures humidity with a mass type sensor. The sensor uses a principle like the piezoelectric effect to measure frequency, force, or strain changes which is super sensitive for mass variation. The piezoelectric dielectric material absorbs water vapor proportionately to the ambient humidity, thus changing the frequency due to the increase of the MEMS resonator device's mass as the device becomes heavier. The humidity changes the mass of the sensor, which is proportional to the relative humidity in the air [19].

3.1.2 Fabrication process of piezoelectric humidity sensor

The most crucial feature of the piezoelectric MEMS resonator is the capability to integrate effectively with other electrical components in semiconductor chips that are deemed as IC-compatible for on-chip applications integrated with IC electronics such as sensing, signal processing, and wireless communication systems. Piezoelectric MEMS resonator is a technology that can be easily fabricated using semiconductor materials on a silicon or silicon on insulator (SOI) wafer substrate and standard fabrication process of material layers such as metal deposition, etching, and patterning [19]. The fabrication process of the piezoelectric mass resonator is described in **Figure 5**.

The fabrication process begins with a photolithography step using a positive photoresist and LOR to define a lift-off profile for the piezoelectric resonator's bottom electrode. After that, a bottom electrode metal layer is deposited following by sputtering the piezoelectric layer (ZnO) with optimized parameters to achieve a (002) c-axis-aligned crystal orientation, as shown in **Figure 5b**. Next, a photoresist is used to pattern the access to define the anchor points for the grounding, which will allow the ohmic contact between the bottom electrode and the subsequent top electrode after etching the ZnO from this anchor point by using ZnO wet etch solution. The top electrode process is performed using LOR and photoresist to have a lift-off profile for defining the piezoelectric resonator's top electrodes following by top electrodes metal layer deposition, as shown in **Figure 5d**. Then, the piezoelectric layer is etched to define the resonator body by performing ZnO dry etch using the DRIE tool for etching the ZnO and the Si anisotropically. Finally, backside etching technique is used to release the device after defining a selected area by patterning a photoresist on the backside of the wafer to allow etching and removal of Si in

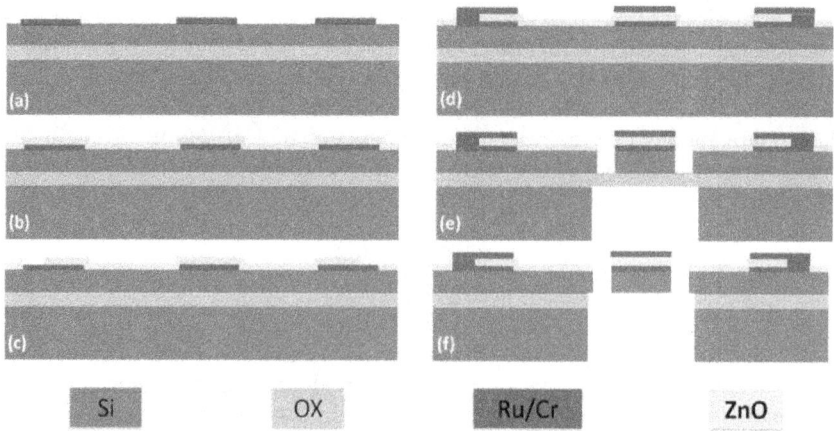

Figure 5.
The cross-sectional view illustrates the fabrication process flow of the ZnO piezoelectric resonator on an SOI wafer including: (a) bottom electrodes patterning; (b) ZnO sputtering; (c) ZnO etch; (d) top electrodes patterning; (e) define the device body; (f) device release [14].

selected areas using HAR DRIE Si etch followed by SiO_2 anisotropic directional dry etch to suspend and fully release the device as shown in **Figure 5f**.

3.2 MOF-based functionalized mass sensors

Metal–Organic Framework (MOF) thin film-coated metal oxide layers have recently been exploited to implement ultra-high sensitivity gas sensors. A thin MOF layer is gradually coated on the ZnO surface to form an ultrahigh sensitive layer to further tune the newly integrated MOF and ZnO materials with desired properties to detect gas and humidity efficiently [20–22]. The piezoelectric sensor can be used in sensing applications based on the changes in the physical properties of the device, such as stress or the mass due to gas absorption at the surface of the device. This device can also be used for humidity sensing applications by coating MOF on ZnO on-Si resonators. The humidity is controlled by detecting the mass changes and monitoring the resonating structure's frequency changes due to the mass loading and surface stress changes (static mode) after absorbing the humidity in the air. Wang's group has grown MOF crystals on ZnO-on-Si resonators to develop a new ultrasensitive sensor for gas and humidity detection by combining the MOF crystals layer, which has excellent absorption and discrimination qualities, and the ZnO layer, which has excellent sensitivity. The sensor can reach a sensitivity of about 191 Hz pg-1after performing FIB- micro pellet depositions and measuring the frequency change per deposition. The frequency change Δf_0 was measured to be 726 Hz [16].

4. Resistive humidity sensing technology

Resistive humidity sensing technology depends highly on the water molecules' absorption into the sensitive material used in the system. Such exposure to humidity can cause either electrical or mechanical effects due to its interaction with the water molecules. Electrical effects, typically impedance change, are measured in standard resistive humidity sensors, while mechanical effects, typically mechanical deformation, are used in piezoresistive humidity sensors. Each of these sensors is designed differently based on the detection mechanisms [23, 24].

4.1 Standard resistive humidity sensors

The standard resistive humidity sensing fabrication process is quite simple, using an insulator material as a starting substrate such as glass. Then, it follows with a metallic patterning of interdigitated electrodes covered by materials that are sensitive to humid environments, as shown in **Figure 6**. The selection of the sensitive materials in these types of sensors determines the quality and the performance of the resistive humidity sensors [23].

The resistive humidity sensors are categorized by the sensitive materials. These materials are generally divided into four groups of materials; ceramics, polymer, electrolytes, and mixer of ceramic/polymer, also known as hybrid composites-based sensors.

Electrolytes based sensors, developed by Dunmore in 1938, are the first developed sensors for humidity detection [26]. The lithium chloride (LiCl) was used by Dunmore as a sensitive material to the humid environment for circuit applications [26]. The humidity can be detected in these types of sensors when the water molecules are absorbed by the electrolyte cells. The electrical effects can also be measured when the cell conductivity changes. The difficulty of working under harsh humid conditions and the low performance had led to developing sensors of other types of materials.

Polymer based humidity sensors are generally categorized into two classes of polymers: polyelectrolytes and conjugated polymers. Typically, polymer humidity sensors' performance depends on the polymer's chemical properties. Polyelectrolytes are hydrophilic and their conductivity is lower than conjugated polymers. Fabrication of polyelectrolytes polymers requires chemical reactions to change the polymers' water solubility without affecting their hydrophilicity, making them inconvenient in contrast with ceramics-based sensors due to their low sensitivity and poor impedance [27]. Several polyelectrolyte materials have been investigated, such as ammonium salt and sulfonate salts [28, 29]. The conductivity in most polymer-based sensors has an inverse relationship with the humidity level. This correlation was observed to be non-linear, unlike piezoresistive-based and capacitive-based humidity sensors [30, 31]. Conjugated polymers are hydrophobic, and their conductivity can be significantly increased by doping metallic ions. By doping the polymer, the humidity decreases at a low level and thus increases the impedance change. Various polymers were doped with different metallic catalysts such as nickel and gold [32, 33]. As a result, the conductivity of the sensors had enhanced as well as other performance parameters such as linearity, sensitivity, response time, etc. [34, 35].

Ceramics based sensors proved to have advantages over polymer-based sensors in terms of mechanical strength, ability to operate at elevated temperature, effectiveness in absorbing water molecules on the surface, and better chemical stability

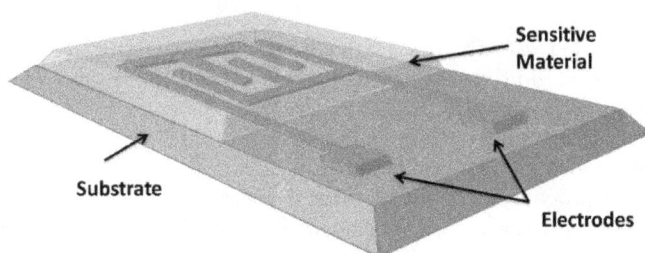

Figure 6.
General schematic of resistive humidity sensors [25].

[27, 28]. Several materials have been studied in literature at different humidity levels and different temperature ranges, such as $MnWO_4$ and SnO_2 [35, 36]. Most of the ceramic based sensors' materials utilize compounded materials to overcome the deficiency issues found in typical materials, such as poor sensitivity and inability to work in harsh environments [12]. On the other hand, ceramic based sensors have incompatibility problems with IC fabrication technology due to their surface contamination [36].

Another type that utilizes the advantages of both polymer sensors and ceramic sensors is the formulation of the composite/compound of the two materials. Recently, such a method has shown its capability to produce sensing elements that offer better performance since they gained the polymer and ceramic materials advantages. Several composites have been investigated in the literature, such as polyaniline and tungsten oxide (PANI/WO_3), TiO_2 nanoparticle/polypyrrole, iron oxide-polypyrrole (Fe_3O_4-PPY) nanocomposite, etc. These hybrid sensors have shown high performance behaviors in response time, mechanical strength, and hysteresis features [37–39].

4.2 Piezoresistive humidity sensors

Unlike standard resistive based humidity sensors where only one material controls the sensor operation of sensing and detection, piezoresistive based humidity sensors require two materials for their operation. As a result, the humidity sensors' performance was improved, especially that they do not exhibit non-linearity problems at low humidity like standard humidity-based sensors [30]. The two materials in these types of sensors are: humid sensitive materials and mechanical sensitive materials or piezoresistive materials. The sensitive materials sense the water molecules and the piezoresistive materials detect the stress changes due to the expansion caused by absorbing the water molecules in the sensitive layer. Piezoresistive based sensors are MEMS devices that are compatible with pre-CMOS and post-CMOS technologies [39]. The fact that the Si has large piezoresistive coefficients has allowed it to be widely used as piezoresistive material over other materials such as metals. It also enabled these sensors to be used in miniaturized devices since Si is the based material in surface and bulk micromachining technologies [39, 40].

The piezoresistive effects were noticed in the 19th century by Wiliam Thomson [41]. He noticed in his experiments that resistivity is related to the mechanical loads in metals, which he used as piezoresistive materials [41]. In the 20th century, the piezoresistive behavior was studied intensively by many scholars in the field and was pioneered by Smith SC [42]. The piezoresistive coefficient in piezoresistive materials relates the resistivity with the mechanical stress as follow:

$$\frac{\Delta\rho}{\rho} = \pi\sigma \tag{8}$$

where, ρ, π, σ are resistivity, piezoresistive coefficient and stress, respectively [24].

Piezoresistive humidity sensors have been developed and miniaturized using micromachining technology. Early designs of piezoresistive sensors were fabricated using bulk micromachining process where SOI wafers are typically utilized as starting wafers and wet bulk etching of the back-side of the wafer leading to some limitation of the process precision. Surface micromachining was also used to develop these types of sensors. J-Q Huang et al. reported a successful method that is compatible with pre-CMOS and post-CMOS technologies using microcantilever as humidity sensors. Such a technique has exhibited better sensors' performance in terms of sensitivity and linearity [39]. A comparison between the three types from

Sensing Technology	Materials	Humidity Range %	Response Time	Sensitivity (/%RH)	Ref.
Capacitive	PI, CU1512, DuPont	30–70	3s	0.86	[43]
	BCBa, 4024–40, Dow Chemical	50–90	0.5s	0.025	[44]
Piezoelectric	ZnO/PI	5–87	50s	34.7	[45]
Resistivity	Si (Piezoresistive)	30–70	1s	4.4	[46]
	PAMPS doped salts	20–90	60s	0.026	[47]
	TiO2 NP3/PPy/PMAPTAC	30–90	30s	0.065	[37]

Table 1.
Comparison of humid sensing technologies.

examples of published work in humid sensing technologies in terms of sensitivity and response time is presented in **Table 1**.

5. Conclusion

This chapter reviewed three types of MEMS humidity sensors: capacitive, piezoelectric, and resistive sensors. While the capacitive sensing depends on the changing permittivity of the sensing material, the humidity can be determined in the piezoelectric sensors by measuring the shift in the resonance frequency. The resistive sensors use the change in resistivity to detect the humidity change. Capacitive sensors in general exhibit higher linearity, faster response and temperature compensation but are sensitive to gas contaminations compared to the resistive sensors [48]. Piezoelectric sensors, on the other hand, do not require external power source which is needed for both capacitive and resistive. The resistive sensors are cheaper to build and have simple readout circuit compared to the other two types [49].

Author details

Ahmad Alfaifi, Adnan Zaman* and Abdulrahman Alsolami
King Abdulaziz City for Science and Technology, Riyadh, Saudi Arabia

*Address all correspondence to: azaman@kacst.edu.sa

IntechOpen

References

[1] Golonka, L. J., Licznerski, B. W., Nitsch, K., & Teterycz, H. "Thick-film humidity sensors." Measurement Science and Technology 1997, 8(1), 92–98. doi:10.1088/0957-0233/8/1/013

[2] Lammel, Gerhard. "The future of MEMS sensors in our connected world." In 2015 28th IEEE International Conference on Micro Electro Mechanical Systems (MEMS), pp. 61-64. IEEE, 2015.DOI: 10.1109/MEMSYS.2015.7050886

[3] Su, P.-G.; Hsu, H.-C.; Liu, C.-Y. "Layer-by-Layer Anchoring of Copolymer of Methyl Methacrylate and [3-(methacrylamino)propyl] Trimethyl Ammonium Chloride to Gold Surface on Flexible Substrate for Sensing Humidity." Sens. Actuators B Chem. 2013, 178, 289–295

[4] Lim, D.-I.; Cha, J.-R.; Gong, M.-S. "Preparation of Flexible Resistive Micro-Humidity Sensors and Their Humidity-Sensing Properties." Sens. Actuators B Chem. 2013, 183, 574–582.

[5] Sadaoka, Y.; Matsuguchi, M.; Sakai, Y.; Aono, H.; Nakayama, S.; Kuroshima, H. "Humidity Sensors Using KH2PO4-Doped Porous (Pb,La)(Zr,Ti)O3." J. Mater. Sci. 1987, 22, 3685–3692.

[6] Wang, J.; Wan, H.; Lin, Q. "Properties of a Nanocrystalline Barium Titanate on Silicon Humidity Sensor." Meas. Sci. Technol. 2003, 14, 172–175

[7] Imran, Z.; Batool, S.S.; Jamil, H.; Usman, M.; Israr-Qadir, M.; Shah, S.H.; Jamil-Rana, S.; Rafiq, M.A.; Hasan, M. M.; Willander, M. "Excellent Humidity Sensing Properties of Cadmium Titanate Nanofibers." Ceram. Int. 2013, 39, 457–462

[8] Korotcenkov, Ghenadii. "Handbook of gas sensor materials." Conventional approaches 1 (2013).https://doi.org/10.1007/978-1-4614-7165-3

[9] Cicek, Paul-Vahé, Tanmoy Saha, BichoyWaguih, Frederic Nabki, and Mourad N. El-Gamal. "Design of a low-cost MEMS monolithically-integrated relative humidity sensor." In 2010 International Conference on Microelectronics, pp. 172-175. IEEE, 2010.DOI: 10.1109/ICM.2010.5696107

[10] Kim, Jihong; Cho, Jang-Hoon; Lee, Hyung-Man; Hong, Sung-Min. "Capacitive Humidity Sensor Based on Carbon Black/Polyimide Composites" Sensors 2021.

[11] Jack R. McGhee, Jagdeep S. Sagu, Darren J. Southee, Peter. S. A. Evans, and K. G. Upul Wijayantha, "Printed, Fully Metal Oxide, Capacitive Humidity Sensors Using Conductive Indium Tin Oxide Inks" ACS Applied Electronic Materials 2020.0222000

[12] Farahani, Hamid, Rahman Wagiran, and Mohd Nizar Hamidon. "Humidity sensors principle, mechanism, and fabrication technologies: a comprehensive review." Sensors 14, no. 5 (2014): 7881-7939.https://doi.org/10.3390/s140507881

[13] Wang, S., Park, S.S., Buru, C.T. et al. Colloidal crystal engineering with metal–organic framework nanoparticles and DNA. Nat Commun 11, 2495 (2020).

[14] Zaman, Adnan, "Hybrid RF Acoustic Resonators and Arrays with Integrated Capacitive and Piezoelectric Transducers" (2020). Graduate Theses and Dissertations.

[15] A. Zaman, A. Alsolami, I. F. Rivera and J. Wang, "Thin-Piezo on Single-Crystal Silicon Reactive Etched RF MEMS Resonators," in IEEE Access,

vol. 8, pp. 139266-139273, 2020, doi: 10.1109/ACCESS.2020.3012520.

[16] James. Williams, Simeon Ochi, and Lewis Research Center. "Characterization of Noncontact Piezoelectric Transducer With Conically Shaped Piezoelement,"Washington, D. C.: National Aeronautics and Space Administration, Scientific and Technical Information Division, 1988.

[17] J. Wang, J. E. Butler, T. Feygelson and C. T. C. Nguyen, "1.51-GHz nanocrystalline diamond micromechanical disk resonator with material-mismatched isolating support," Proceedings of 17th IEEE International Conference on Micro Electro Mechanical Systems (MEMS 2004), pp. 641-644, 2004.

[18] M. Mahdavi, E. Mehdizadeh, S. Pourkamali, "Piezoelectric MEMS resonant dew point meters," Sensors and Actuators A: Physical, Volume 276, 2018, Pages 52-61, ISSN 0924-4247.

[19] Rivera, Ivan, "RF MEMS Resonators for Mass Sensing Applications" (2015). Graduate Theses and Dissertations.

[20] I . Rivera, A. Avila, J. Wang, "Fourth-Order Contour Mode ZnO-on-SOI Disk Resonators for Mass Sensing Applications," Actuators 2015, 4, 60–76

[21] Ahmad, Waqas et al. "Highly Sensitive Humidity Sensors Based on Polyethylene Oxide/CuO/Multi Walled Carbon Nanotubes Composite Nanofibers." Materials (Basel, Switzerland) vol. 14,4 1037, 2021.

[22] Balasubramanian, S, Polaki, S, Prabakar, K, "Ultrahigh sensitive and ultrafast relative humidity sensing using surface enhanced microcantilevers" Smart Materials and Structures, *et al* 2020.

[23] Tang QY, Chan YC, Zhang K "Fast response resistive humidity sensitivity

of polyimide/multiwall carbon nanotube composite films." Sensors Actuators B Chem 2011, 152(1):99–106. https://doi. org/10.1016/j.snb.2010.09.016

[24] Kloeck, B. (1993). Piezoresistive Sensors. In Sensors (eds H.H. Bau, N.F. deRooij and B. Kloeck)

[25] Cha, Jae-Ryung & Gong, Myoung-Seon. "AC Complex Impedance Study on the Resistive Humidity Sensors with Ammonium Salt-Containing Polyelectrolyte using a Different Electrode Pattern." Bulletin of the Korean Chemical Society, 2013. 34. 10.5012/bkcs.2013.34.9.2781.

[26] Dunmore, F. "An Electric Hygrometer and Its Application to Radio Meteorography." J. Res. Natl. Bur. Stand. 1938, 20, 723–744

[27] Sakai, Y.; Sadaoka, Y.; Matsuguchi, M. "Humidity Sensors Based on Polymer Thin Films." Sens. Actuators B Chem. 1996, 35, 85–90.

[28] SU, P., & UEN, C. "A resistive-type humidity sensor using composite films prepared from poly(2-acrylamido-2-methylpropane sulfonate) and dispersed organic silicon sol." Talanta, 2005, 66 (5), 1247–1253. doi:10.1016/j. talanta.2005.01.039

[29] Gerlach, G., & Sager, K. "A piezoresistive humidity sensor." Sensors and Actuators A: Physical, 1994, 43 (1-3), 181–184. doi:10.1016/0924-4247 (93)00690-6

[30] Lee, H.; Jung, S.; Kim, H.; Lee, J. "High-Performance Humidity Sensor with Polyimide Nano-Grass." In Proceedings of 2009 International Solid-State Sensors, Actuators and Microsystems Conference (TRANSDUCERS 2009), Denver, CO, USA, 21–25 June 2009; pp. 1011–1014

[31] Yang, M.; Li, Y.; Zhan, X.; Ling, M. "A Novel Resistive-Type Humidity

Sensor Based on Poly(p-Diethynylbenzene)." J. Appl. Polym. Sci. 1999, 74, 2010–2015.

[32] Su, P.-G.; Shiu, C.-C. "Electrical and Sensing Properties of a Flexible Humidity Sensor Made of Polyamidoamine Dendrimer-Au Nanoparticles." Sens. Actuators B Chem. 2012, 165, 151–156.

[33] W. Qu, J. U. Meyer, "Thick Film Humidity Sensors Based on Porous MnWO4 Material," Measurement Science and Technology, Vol. 8, pp. 593-600, 1997.

[34] Yamamoto, T.; Shimizu, H. "Some Considerations on Stability of Electrical Resistance of the TiO2/SnO2 Ceramic Moisture Sensor." IEEE Trans. Compon. Hybrids Manuf. Technol. 1982, 5, 238–241

[35] Traversa, E. "Ceramic sensors for humidity detection: the state-of-the-art and future developments." Sensors and Actuators B: Chemical, 23(2-3), 135–156. 1995. doi:10.1016/0925-4005(94)01268-m

[36] Parvatikar, N.; Jain, S.; Khasim, S.; Revansiddappa, M.; Bhoraskar, S.V.; Ambika Prasad, M.V.N. "Electrical and Humidity Sensing Properties of polyaniline/WO3 Composites." Sens. Actuators B Chem. 2006, 114, 599–603.

[37] Su, P.-G.; Wang, C.-P., "Flexible Humidity Sensor Based on TiO2 Nanoparticles-Polypyrrole-Poly-[3-(methacrylamino)propyl] Trimethyl Ammonium Chloride Composite Materials." Sens. Actuators B Chem. 2008, 129, 538–543.

[38] Tandon, R.P.; Tripathy, M.R.; Arora, A.K.; Hotchandani, S., "Gas and Humidity Response of Iron oxide—Polypyrrole Nanocomposites." Sens. Actuators B Chem. 2006, 114, 768–773.

[39] Huang, J.-Q.; Li, F.; Zhao, M.; Wang, K. "A Surface Micromachined

CMOS MEMS Humidity Sensor." Micromachines 2015, 6,1569-1576. https://doi.org/10.3390/mi6101440

[40] H. Hammouche, H. Achour, S. Makhlouf, A Chaouchi, M. Laghrouche, "A comparative study of capacitive humidity sensor based on keratin film, keratin/graphene oxide, and keratin/carbon fibers", Sensors and Actuators A: Physical, 2021

[41] Thomson W., "On the electro-dynamic qualities of metals: – effects of magnetization on the electric conductivity of nickel and of iron." In: Proceedings of the Royal Society of London. JSTOR, pp 546–550, 1856.

[42] Smith CS, "Piezoresistance effect in germanium and silicon." Phys Rev 94 (1):42–49, 1954, https://doi.org/10.1103/PhysRev.94.42

[43] M. Dokmeci, K. Najafi, "A high-sensitivity polyimide capacitive relative humidity sensor formonitoring anodically bonded hermeticmicropackages," J. Microelectromech. Syst. 10 (2) 197–204, 2001.

[44] A. Tetelin, C. Pellet, C. Laville, G. N'Kaoua, Fast response humidity sensors for a medical microsystem, Sens. Actuators B 91 (1–3) 211–218, 2003.

[45] He X. L., Li D. J., Zhou J., Wang W. B., Xuan W. P., Dong S. R., Jina H. and K. L. J., "High sensitivity humidity sensors using flexible surface acoustic wave devices made on nanocrystalline ZnO/polyimide substrates" J. Mater. Chem., 2013

[46] Xu J., Bertke M., Li X., Mu H., Zhou H., Yu F., Peiner E., "Fabrication of ZnO nanorods and Chitosan@ZnO nanorods on MEMS piezoresistive self-actuating silicon microcantilever for humidity sensing." Sensors and Actuators B: Chemical, 273, 276–287. doi:10.1016/j.snb.2018.06.017, 2018

[47] P.G. Su, W.C. Li, J.Y. Tseng, C.J. Ho, "Fully transparent and flexible humidity sensors fabricated by layer-by-layer self-assembly of thin film of poly(2-acrylamido-2-methylpropane sulfonate) and its salt complex," Sens. Actuators B 153 29–36, 2011.

[48] Yadav, Anuradha. "Classification and applications of humidity sensors: a review." Int. J. Res. Appl. Sci. Eng. Technol. 6, no. 4 (2018): 3686-3699.

[49] Harrey, P. M., B. J. Ramsey, P. S. A. Evans, and D. J. Harrison. "Capacitive-type humidity sensors fabricated using the offset lithographic printing process." Sensors and Actuators B: Chemical 87, no. 2 (2002): 226-232.

Chapter 5

Graphene and Its Nanocomposites Based Humidity Sensors: Recent Trends and Challenges

Avik Sett, Kunal Biswas, Santanab Majumder,
Arkaprava Datta and Tarun Kanti Bhattacharyya

Abstract

Humidity sensors are of utmost importance in certain areas of life, in processing industries, in fabrication laboratories and in agriculture. Precise evaluation of humidity percentage in air is the need of various applications. Graphene and its composites have shown great potential in performing as humidity sensors owing to enormous surface area, very low electrical noise, high electrical conductivity, mechanical and thermal stability and high room temperature mobility. There is no such extensive review on graphene-based devices for humidity sensing applications. This review extensively discusses graphene-based devices intended towards sensing humidity, starting from the methods of synthesizing graphene, its electronic and mechanical properties favoring sensing behavior and different types of sensing mechanisms. The review also studies the performance and recent trends in humidity sensor based on graphene, graphene quantum dots, graphene oxide, reduced graphene oxide and various composite materials based on graphene such as graphene/polymer, graphene/metal oxide or graphene/metal. Discussions on the limitations and challenges of the graphene-based humidity sensors along with its future trends are made.

Keywords: Humidity sensors, Synthesis of graphene, Graphene oxide, Graphene quantum dots, Graphene/2D materials

1. Introduction

Humidity sensors is one of the habitual sensors used in our regular activities. It plays a significant part in controlling humidity in certain application areas such as monitoring moisture in agriculture, processing industry, device fabrication laboratories, monitoring indoor air quality, controlling domestic instruments, medical applications and in weather forecasting [1, 2]. Humidity sensors detects the amount of water in the atmosphere and transforms it into a measurable signal. After interaction with the water molecules, the physical parameters of the humidity sensors are observed to change. These parameters decide on the category of humidity sensor such as resistive, capacitive, optical-fiber, impedance, surface-acoustic wave (SAW), Quartz crystal microbalance (QCM) and resonance type [3]. Many ultra-sensitive materials have been developed towards detection of water molecules such as ceramics (Al2O3, SiO2 etc.) [2], perovskite compounds [4, 5], semiconductors

(SnO2, ZnO, etc.) [6, 7], semiconducting and conducting polymers [8] and 2-D materials such as MoS2 [9], WS2 [10], black phosphorus [11], carbon nanotubes [12] and graphene [13].

Graphene and graphene based composite nanomaterials in this context is attracting significant attention due to its fascinating and unique electrical, mechanical and optical properties. Various methods of graphene synthesis have been discovered such as sublimation of silicon carbide (SiC) at high temperature, chemical vapor deposition (CVD), mechanical exfoliation etc. However, to ensure the availability of graphene-based sensor devices widely to the society, cost of these sensors must be as low as possible and large-scale production of graphene should be feasible. Production of chemically derived graphene by reducing graphene oxide is an economical process and requires very user friendly and low-cost equipment and resources. Modified Hummers method-based synthesis of graphene is well known. Various reducing techniques of graphene oxide leads to the formation of chemically derived graphene (reduced graphene oxide) with different functional groups attached to it. Chemically derived graphene induces several defects on its matrix compared to pristine graphene. These defect sites act as electroactive and optical centers for sensing applications. Tuning the mechanical, optical and electrical properties of graphene for various sensing applications can be done quite efficiently by i) metal doping, ii) decorating it with nanoparticles, iii) selectively reducing the functional groups, iv) making nanocomposites with metal oxide nanoparticles or polymers and many more.

Even though there are some articles related to progress of chemical, tactile and gas sensors based on graphene [14, 15], there are no such review article on humidity sensors based on graphene. The main focus of this review is to discuss the recent advances in humidity sensors based on graphene. The review is partitioned into three parts: Synthesis of graphene and its related properties favoring sensing phenomenon, sensing mechanism of different humidity sensors and advances made in graphene and its composite towards fabrication of humidity sensors. The synthesis part explains the different methods of preparing pristine graphene, chemically derived graphene and also describes the different properties of graphene that facilitates the use of graphene as a sensing material. The different mechanisms include seven types of graphene-based humidity sensors and its description. The last part discusses the progress made in humidity sensor development using graphene and graphene-based composites as sensing material. Finally, challenges faced by graphene-based humidity sensors and the outlook towards future research and developments are discussed.

2. Graphene as a transducing material owing to its properties

Graphene is a sp2 hybridized honeycomb like structure with an array of carbon atoms featuring a bond length of 1.42 Å and 120° angles. It is a monolayer structure being established as the thinnest form of carbon nanomaterials and probably the thinnest material discovered. The symmetry of this two-dimensional lattice gives rise to some fascinating electronic and mechanical properties. Graphene is known to be the strongest material ever found [16] with an ability to be bent without affecting its internal structure and properties. Graphite being the parent material, comprises stacking of several graphene layers on top of one another having an interlayer separation of 3.354 Å [17]. There are two different ways in which the graphene layers could be stacked as shown in **Figure 1**. Eighty percent of the graphite found in nature is the Bernal graphite, an allotrope of graphite comprising ABA stacking configuration. Fourteen percent of the graphite found is of the type rhombohedral

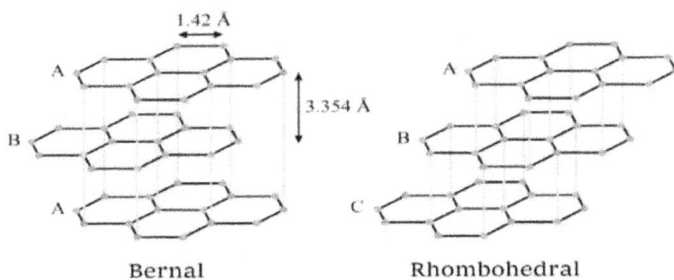

Figure 1.
Bernal (ABA) and rhombohedral (ABC): The two stable allotropes of graphite.

with ABC stacking configuration, whereas the rest six percent is disordered in nature having no such stacking configuration. Single layer graphene is very difficult to obtain and it has very low yield. Hence few layered graphene is considered in most of the cases for device fabrication towards heavy metal sensors, as its properties are in between that of graphite and single layer graphene, with high mechanical and electrical stability. However, both single layer and few layered graphene can be used towards sensing application after proper functionalization.

2.1 Electronic properties

The formation of perfect honeycomb lattice from the arrangement of carbon atoms results in electronic properties of graphene. Hybridized orbitals that are sp2 in nature is the backbone of graphene. The p orbitals that are perpendicular to the lattice is responsible for conductivity in the graphene matrix. These orbitals get conjugated to form conduction bands (π^*) and valence bands (π), which are significant towards sensing of various analytes. The transduction of the interaction between the graphene-based materials and the analytes to a readable electrical signal is facilitated by these orbitals. The fermi level of pristine graphene is located exactly in the position where the conduction and valence bands meet. Altogether, in the Brillouin zone, there are six such points where conduction band meets valence band. These points are termed as neutrality points or Dirac points and in the reciprocal space, it is labeled as K and K' [18]. The ability of electrons in graphene to travel sub-micron distances without any scattering (travel without electrical resistance) is known as ballistic transport. This type of transport in graphene is observed at room temperature and is not affected by the adsorbates present in the graphene matrix or the substrate's topography [19]. Even though at room temperature, impurity scattering limits the mobility of charge carriers, the mobility of carriers in graphene stays very high even at the presence of chemical and electrical dopants at high concentrations [20]. Electrical robustness of graphene-based sensors is attributed to these types of transport phenomenon. Tuning of charge carriers between holes and electrons for doping graphene is possible due to the presence of strong ambipolar electric field effect. This facile modulation of graphene field effect transistors allows detection of n-type and p-type analytes with very high sensitivity, as will be discussed later. At the fermi level, the conductivity of graphene does not get vanished as conductivity of graphene is quantized. Eventually, at the fermi level where density of states approaches zero, there still exists finite conductivity. The last charge carrier provides e2/h as the minimum conductivity [18]. Hence, graphene is found to have no transition of metal–insulator while measurement of σ_{min}. The experimentally obtained value for σ_{min} is found to be 4e2/h. The electronic property of graphene changes with change in the number of graphene layers. At one horizon, single layer graphene is a semiconductor having zero band gap (semimetal having zero overlap) and on the other horizon, graphite is found to be a semimetal having a band overlap of 41 meV. With respect to the properties of graphene, the threshold between bulk and single layer graphene is not much clear. With increase in the number of graphene layers, the band overlap starts increasing till it reaches that of the bulk graphite. The amount of overlap seems to be a decent indicator of the threshold between 2D and 3D behavior of graphene. It is observed that, for graphene with more than 11 layers, the overlap is seen to vary only 10% compared to that of bulk graphite. Hence, 10 layers are suggested to be the limit to few layers' graphene. Therefore, single layer graphene as well as few layers graphene (up to 10 layers) can be used towards water contaminant and heavy metal detection.

2.2 Mechanical properties

Molecular dynamics-based simulations are carried out to investigate the fracture strength and Young's modulus of single layer graphene. Measurements related to force-displacements by AFM (atomic force microscopy) was conducted to analyze the Young's modulus of few layer graphene. The AFM measurements were conducted on strips of graphene after being suspended over trenches. Measurements using AFM by nanoindentation was conducted to measure the intrinsic breaking strength and other elastic properties of graphene [21]. Defect free single layer graphene was found to have fracture strength of 130 GPa and Young's modulus of 1.0 TPa.

3. Synthesis of graphene for humidity sensors

Synthesis of graphene can be categorized into two different approaches, top-down and bottom-up. Top-down approaches deal with exfoliation of bulk graphite into single layer (or few layer) graphene. Bottom-up approaches include pyrolysis, epitaxial growth, chemical vapor deposition and plasma synthesis.

3.1 Top-down techniques

The top-down technique deals with attacking the raw bulk graphite and separate the layers to obtain sheets of graphene. Mechanical and chemical exfoliation of bulk graphite powder results in single to few layers of graphene sheet characterized by Raman spectroscopy.

3.1.1 Mechanical exfoliation

Exfoliation of bulk graphite mechanically to extract single layer to few layer graphene sheets over desired substrates is a well-known method. The superficial part of the graphite experiences longitudinal or transverse stress during mechanical exfoliation. The interlayer distance between graphene layers is 3.354 Å and the value of bond energy is found to be 2 eV/nm2. The external force required for mechanical cleaving of graphite layers to obtain single atomic layer graphene sheet is approximately 300 nN/mm2. When the partially filled p-orbitals overlap perpendicularly on a plane sheet, the van der Waals forces results in stacking of the sheets. The higher lattice spacing in vertical direction and poor bonding forces between the layers causes the reverse step of stacking, exfoliation to occur when external forces are applied. This mechanical exfoliation is carried out by peeling layers from different graphitic substances such as HOPG (highly ordered pyrolytic graphite), natural graphite and monocrystal graphite. Mechanical exfoliation is generally carried out using different agents such as ultrasonication, electric field and scotch tape.

3.1.2 Chemical reduction of graphite oxide

One of the top-down approaches to produce enormous amount of graphene is through chemical reduction of graphite oxide. Synthesizing graphite oxide requires oxidation of graphite, which is usually done by utilizing certain oxidants such as potassium permanganate, concentrated nitric acid and sulfuric acid. In 1860, graphene oxide was first produced by Brodie, Staudenmaier and Hummers. After this discovery, modified Hummers method and Improved Hummers method were created [22]. Major differences between these methods are the types of oxidants, the number of steps and degree of toxicity. With progress, the toxic nature of synthesis could be eliminated with decrease in reduction steps. Once the oxidation is complete, various reducing agents are employed to remove the oxidizing groups from graphene oxide. The reducing agents may be sodium borohydrate, hydroxylamine, glucose, pyrrole, ascorbic acid, phenyl hydrazine and many more. Formation of reduced graphene oxide suspensions in organic solvents is of common practice, because graphene oxide is hydrophilic in nature, but reduced graphene oxide is hydrophobic, which leads to agglomeration of graphene sheets. Chemical reduction of graphene oxide is a very famous technique among the research community due to i) large quantity of graphene sheets, ii) desired functionalization of graphene sheets for various applications including sensing; transparent conducting electrodes for solar cell; electrochemical high energy electrode material for energy storage and many more; and iii) tuning defects or presence of functional groups by various approaches of reduction.

3.2 Bottom-up techniques

Bottom-up techniques deal with formation of graphene from carbonaceous species in vapor state. The formation of graphene from atoms of carbon results in defect free pure graphene sheets. However, the yield of graphene which is low in such cases, should be improved. Bottom-up strategies include pyrolysis, epitaxial growth, CVD, plasma synthesis and many more.

3.2.1 Pyrolysis

The pyrolysis technique comprises of formation of graphene sheets chemically by solvo-thermal method. For example, one may use 1:1 molar ratio of sodium and ethanol in a container vessel during the entire duration of thermal treatment. Another approach is pyrolization of sodium ethoxide through the process of sonication. This results in better performance towards detachment of graphene sheets.

3.2.2 Epitaxial growth

One of the methods to grow graphene on substrate surfaces is epitaxial growth. Graphene can be prepared by heating and cooling down a crystal of silicon carbide (SiC) at an optimum pressure. Graphene can grow on both silicon and carbide face. Growth over silicon face of the crystal leads to the formation of monolayer or bi-layer graphene. However, growth on the carbide face of the crystal leads to the formation of few layer graphene [23]. The parameters used during the growth process i.e. temperature, heating rate (ramp) and pressure facilitates the quality and nature of graphene produced. In case, high temperature and pressure is applied, growth of carbon nanotubes is favored compared to graphene. Hence, by varying the conditions, different morphology and materials can be obtained such as carbon nanotube, graphene sheets, graphene nanoribbons etc. The lattice of Ni(III) surface is very close to that of graphene with a difference of only 1.3%. Hence, by approach of nickel diffusion [24], a thin layer of nickel might be evaporated onto the SiC crystal surface. After heat is applied, graphene will be found on the surface due to diffusion of carbon through the nickel layer. However, the entire process depends on the temperature and heating rate. This additional layer of nickel allows better separation of graphene grown on the surface of silicon carbide crystal. However, due to the formation of defects and grain boundaries, the quality of graphene is not the best and the as formed graphene lacks homogeneity.

3.2.3 Chemical vapor deposition (CVD)

CVD is one of the best techniques to produce high quality graphene sheets. It is basically a technique which involves reaction in the gaseous phase and deposition onto a desired substrate [25]. The combination of gaseous species in a reaction chamber occurs at a desired optimum pressure and temperature. When the reactant gases meet the substrate in a heated reaction chamber, a material film is seen to grow over the substrate. There are certain by products in the chamber, which are removed during the process. The temperature of the substrate plays a significant role to facilitate proper reaction mechanisms. The speed of deposition of material over the substrate is very low and in the range of few microns per hour. Two types of CVD are widely used, namely, ultra-high vacuum chemical vapor deposition (UHVCVD) and low-pressure chemical vapor deposition (LPCVD).

4. Graphene based humidity sensing mechanism

In this section, various varieties of graphene-based humidity sensing mechanism have been explored, which identifies the intrinsic mechanism involved with the water molecules in connection with the graphene's surface architecture. The interface between the available water molecules over the surface of graphene material makes a net deflection in the signal response making it an ultrasensitive sensing material for humidity sensing approach. This section highlights five types of sensing mechanisms for graphene-based humidity sensors, which includes the SAW, FET, resistive, capacitive and optical fiber-based sensing mechanisms.

4.1 Humidity sensors based on field-effect transistor (FET)

Owing to the ease of miniaturization, nature of portability, higher sensitivity, FET based Gas sensing has been designed for widespread applications [26]. Typically, a graphene-based humidity sensor comprises of graphene as the channel material, a thin dielectric layered gate electrode (**Figure 2**) and source and drain electrodes, respectively. It is through the dielectric layer; a bias voltage is usually applied on the gate electrode which has the role to modulate the graphene channel's intrinsic conductivity. It is under a constant applied gate voltage; the net humidity sensing is usually measured by measuring the current change of the graphene channel before and after exposing the channel to humidity environment. Usually, the underlying mechanisms associated with the change in the current response lies in the fact that whenever the water or gaseous molecules comes into the closer

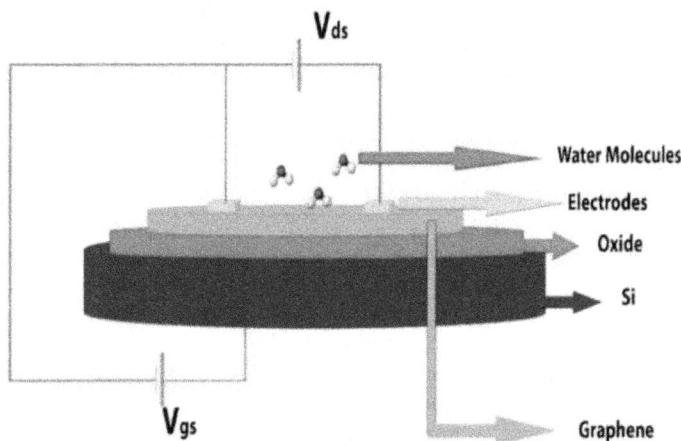

Figure 2.
Schematic of graphene field effect transistor.

proximity of graphene's surface molecules, there is change and alterations in the intrinsic electronic structure of the sensing material. In our case, the sensing material graphene is a material of augmented surface area (theoretical surface area of 2630 m2/gm). Such a huge surface area makes graphene an ideal candidate for functionalization with several functional groups for making the humidity sensors sensitive and much more robust. It is usually noticed that gases like CO, H2, SO2, H2S, NH3, NO, NO2 and ethanol can be detected in a sensitive manner owing to the gaseous molecular interaction between the exposed gaseous molecules and the graphene sensing material [27].

4.2 Resistive humidity sensors

Resistive humidity sensing has become one of the widest used humidity sensing platforms because of its intrinsic low driven power, easy miniaturization, higher rate of reusability and simplicity in operation and that too at the economical scale. In general, the change in humidity sensing is based on the net change in the electrical resistance of the graphene sensing material, which gets changed with the adsorbing water molecules over the sensing material (**Figure 2**). The configuration of the typical resistive sensor is based on the two conductive electrodes placed on an inert substrate, which besides the resistance change for humidity change can detect the net current change of the sensing mechanism. In order to augment the sensing area of the material, the employed electrodes are usually interdigitated. In the underlying principle, the sensing materials can be made further sensitive by coating the sensing materials with different functional groups, making more surface area available for adsorption of exposed water molecules, making porous structure, compositing the sensitive materials with other available more sensitive components. In this category of sensors, different gaseous analytes like CO, H2, NO2, Cl2, O2 and other organic vapors can be detected by changing the different sensing materials [28–30].

4.3 Capacitive based humidity sensors

Humidity sensors based upon capacitive properties of the GO materials lies in the fact that GO based sensing materials are insulating in nature. It is seen that with the available water molecules adsorbed over the GO structure, there is a proportional increase in the proton conductivity rate. This phenomenon signifies the rate of water detection over the structure of GO when tested using capacitive mode. Upon water molecule adsorption, there is proton conductivity which results into change in the capacitance value [31]. In this view it could be argued that GO based capacitive sensors, typically works as water-sensitive dielectric in a purely double-layered electrochemical capacitor for the slightest detection of the humidity available in the surrounding atmosphere. In the configuration side, capacitive based humidity sensors possess the in-plane type to expose the maximum active surface area and is consists of two portions, GO-based sensing materials as the dielectrics and the interdigitated conductive electrodes [32]. In the underlying architecture of the capacitive based humidity sensors, the interdigitated conductive electrodes can be made up of metal materials, as well as rGO based coating for all humidity-based sensing approach. In order to make the sensing material of GO a dielectric based humidity sensing, GO can be composited with the different proton conductive material for making the GO based capacitive based humidity sensing sensitive and robust in nature.

4.4 Surface acoustic wave (SAW) based humidity sensors

The Surface Acoustic Wave (SAW) humidity sensors are basically micro-electromechanical systems, whose detection of humidity is solely based upon the modulation of the surface acoustic waves. Typically, a SAW based humidity sensor comprises of a decay line in between the two interdigitated transducer (IDT), piezoelectric substrate [33]. Mechanical SAW is produced with the input IDT signal under the sinusoidal electrical signal, which uses the net piezoelectric effect of the utilized piezoelectric substrate. The surrounding humidity environment influences the obtained SAW signal. Electrical signal is thus been produced by the available SAW signal by the piezoelectric effect. It is perceived that the changing signals of the amplitude, phase, frequency or time delay could be associated with the occurrence in the sensing phenomenon. The change in the physical properties of the sensing materials like mass, visco-elasticity, conductance also affects the acoustic wave velocity or the net attenuation [34]. It is understood that graphene-based SAW detection of humidity sensing lies in the underlying principle that hydrophilic GO or GO based composites makes the mass dependence SAW signaling of humidity detection. The GO and GO based sensing material consist of defects, which entrap the water molecules for making the mass dependence of SAW, contributing to sensing phenomenon of humidity sensors. It has been investigated by Balashov et al., by preparing a type of humidity sensors by coating a submicron-thick film of GO or a PVA thin film on the decay line, and the same GO based humidity sensor exhibits a sensitivity of 1.54 kHz/ % RH, which is much higher than the 0.47 KHz/% RH and 0.13 KHz/ % RH for the PVA coated and uncoated one [33]. A higher sensitivity from 0.5% RH to 85% RH has been shown by few studies, whereby ZnO piezoelectric thin film has been grown on the glass substrate with GO as the sensing material and layer. The set-up has showed the fastest response (from rise time 1 s and fall time of ~19 s) [35]. Subsequent studies have also revealed that flexible SAW humidity sensors with a different substrate material of polyamide substrate [36]. There are also reports whereby SAW humidity sensors were designed based on the GO film/ ZnO/Si substrate. The set up leads to achieved sensitivity of ~91 KHz/% RH with a net linear response towards humidity sensing in the working range of 20–98% [37]. Another study has revealed that AlN/Si doped layered structure work as SAW humidity sensor, whereby GO has been used as the base sensing material along with the low temperature coefficient of frequency. The value of 42.08 KHz/% RH in RH higher than 80% has worked excellent at both the lower value of 90% RH humidity showing an improved thermal stability and outstanding short-term repeatability.

4.5 Optical fiber-based humidity sensors

Optical fiber-based humidity sensors work on the principle of sensing humidity inside the optical fiber which gets sensed by deflecting the water molecules inside the optical fibers, dielectric power or the refractive index. As compared to the conventional electronic humidity sensors, such types of sensors have an added advantage of working in the harsher environments, which comprises the flammable environments, higher temperature and electromagnetic immunity. Also, it has been studied that owing to the higher cost of fabrication repeatability, higher cost of optical equipment's has resulted into the limitations in the applications of optical fiber-based humidity sensors. Owing to the excellent water-adsorbing properties of the GO films, graphene materials have been selected as the potential candidate for the humidity sensing inside the optical fiber which has become the best platform for preserving data loss and accurate humidity sensing paradigm. A power variation of up to 6.9 dB in the higher relative humidity range (70–95%) has been achieved by Xiao et al. who have shown the RGO coated side-polished fiber. Such study has shown a correlation efficiency of ~98.2%, a sensitivity of 0.31 dB/% RH, a response speed of faster than 0.13% RH/s with an overall good rate of repeatability in the 75–95% RH [38].

They have shown that the humidity sensor-based on side-polished single mode fiber coated with GO films, resulted into higher and better performance. The polymer channel waveguide upon subjection of the TE mode adsorption at 1550 nm gets influenced by the water molecular adsorption over the surface of GO, which contributes to the linkage between the optical absorption and the net relative humidity mechanism. One of the research studies have shown that Bragg grating coated GO films exhibited a maximum humidity response with sensitivity ~0.129 dB/ % RH which consists a linear coefficient of 99% proportion under the range of 10–80%. This is further dependent upon the water molecular adsorption and desorption phenomenon over the GO films. It has been also shown by them that the optical fiber humidity sensors based on in-fiber Mach-Zehnder interferometer coated with GO or GO/PVA composite shows linearity and good stability [39]. Several studies have highlighted the deposition of Fabry-Perot resonator humidity sensors which is shown by depositing the RGO films or graphene quantum dots [40, 41] over the surface of the hollow core fiber, along with slight leakage of resonant wavelength. Refractive index of the material plays a paramount role in determining the type of RGO based optical sensor in this case. It has been studied that optical fibers coated with the graphene-based materials not only has been used in the humidity sensors but also been used in the biological and chemical sensors.

5. Different graphene-based materials for humidity sensing

It is well known that the molecules of water present in the environment ensures electronic and physical changes in the sensing materials based on graphene. It is noted that by monitoring different property changes such as resistance, mass, surface acoustic wave, capacitance and impedance, detection of humidity is possible. The current section elaborates recent trends in humidity sensors based on graphene. The following section describes sensors based on electronic properties utilizing technique of measurement such as resistance, capacitance and impedance. Six sub classes of graphene-based material are studied in the following section.

5.1 Pristine graphene-based humidity sensor

FET based gas sensor with pristine graphene as sensing material was first demonstrated by Novoselov et al. Micromechanical cleavage of graphite led to formation of pristine graphene which was transferred to the surface of oxidized wafers of silicon. This layer of graphene not only responded to water molecules but also detected presence of ammonia and oxides of carbon and nitrogen [26]. The electronic properties of the gas sensor induce separate responses towards different gases, like ammonia and carbon monoxide act as donors whereas nitrogen dioxide and humidity act as acceptors. Exceptional high sensitivity of graphene-based sensors is achieved due to extremely low noise of graphene. The electronic property of graphene such as band gap varies with change in humidity environment. The band gap is seen to increase with increase in humidity due to increase in adsorbed water molecules on the surface of graphene. This leads to increase in resistivity with increase in humidity [42]. Smith et al. fabricated graphene by CVD process over silicon dioxide and demonstrated an efficient humidity sensor based on resistance change mechanism. The sensor showed fast response and recovery of less than 1 sec due to quick physisorption and desorption of water vapor over the surface of graphene. Simulation showed that the interaction of impurity bands of the substrate with the electrostatic dipole moment of the water molecules leads to sensitivity of the graphene surface to humidity. Popov et al. reported an interesting phenomenon that the adsorption of water molecules at different locations lead to variable resistance change due to divergent mechanism of interaction. An increase in resistivity is observed when the water molecules are adsorbed at the grain boundary defects which occurs due to the p-type nature of graphene and donor type behavior of the water molecules. However, adsorption of water molecules at the edge defects of multilayer graphene contributes to ionic conductivity due to formation of conductive chains. A study by Son et al. revealed that the humidity sensing performance hardly undergoes a change due to physical defects whereas chemical defects may contribute to enhancement in performance. This can be achieved by thickness control of the graphene layer and the area of coverage with PMMA over the graphene surface [43].

A bi-layer graphene-based gas sensor was developed by Fan et al. [44], and its investigation showed a quick humidity response and recovery of the sensor. Experimental calculations supported by theoretical analysis showed that the response of bi-layer graphene was less compared to single layer graphene. Zhu et al. developed a highly efficient humidity sensor, however the sensing material used was wrinkled graphene [45]. The wrinkles prevented aggregation of water molecules (microdroplets) over the graphene surface which facilitated fast desorption leading to quick recovery. The sensor is observed to show ultrafast response of 12.5 ms and can be utilized in monitoring sudden respiratory rate and depth changes. CVD grown graphene over woven fabrics was also developed which was used to detect humidity and temperature simultaneously. A study [46] revealed that interlayer interaction of graphene gets modified when the humidity is over 50%. This calls for further attention to stability and recoverability of multi-layer graphene along with focus on repeatability of the sensor. Hence, even with quick response and recovery, the stability and sensitivity of pristine graphene is an area of improvement.

5.2 Graphene oxide-based humidity sensor

Large scalability, very low cost, high yield, simple preparation technique and high sensitivity to proton-conductivity has encouraged investigation of humidity sensors based on graphene oxide (GO). The proton conductivity helps in detection of water molecules through impedance or capacitance signals [47]. A humidity sensor was constructed by Bi et al. where they deposited GO over interdigitated micro-electrodes and measured the change in capacitance with a LCR meter. The change is capacitance was observed with change in humidity levels and it also varied with frequency. The sensitivity of the device was found to be high in the RH range 15–95%. The sensor exhibited a response time of 10.5 sec and recovery time of 41 sec. Park et al. demonstrated the relation between adsorption/desorption hysteresis with sensitivity [48] of GO film working in conductometric mode. The study revealed that sensors fabricated at pH 3.3 shows reduced sensitivity and less hysteresis error than the sensors fabricated at pH 9.5. Many strategies are taken up to improve the humidity sensing performance of the GO films such as ultra-large size of GO to enhance the proton conductivity, doping of GO with heteroatoms, GO foams that are free standing facilitating increase in active sites, GO coated over silk fiber to benefit flexibility of the sensor and many more. Graphene oxide may be coated over a flexible substrate to function as a wearable colorimetric humidity sensor as demonstrated by Hong et al. [49].

In general, interdigitated electrodes are prepared by photolithography to contain the GO sensing layer for humidity sensing application. However, gold and silver are quite expensive and patterning requires expensive instruments, skilled professionals and complex procedures. Nowadays, direct laser writing techniques are used to develop electrodes in non-contact mode, with no need for post processing, clean room and are very compatible with commercial electronic product lines such as sensors, devices for energy storage and self-powered devices [50]. Laser irradiation may be used to reduce graphene oxide to form conductive electrodes of reduced graphene oxide with very high conductivity. However, graphene oxide may still act as humidity sensitive sensing layer in the same device. Ajayan et al. developed a micro-supercapacitor (interdigitated) over a hydrated graphene oxide film and demonstrated that proton conductivity in graphene oxide is directly dependent on humidity concentration. A facile fabrication of humidity sensor based on rGO/GO/rGO was reported over a flexible substrate of PET, which used direct laser writing with semiconductor diode laser. The black color of GO turned to gray rGO when laser was irradiated over the GO surface with evolution of certain gases. The rGO so formed was characterized by Raman spectroscopy, X-ray diffraction and X-ray photoelectron spectroscopy. The performance of GO based humidity sensors when evaluated shows very fast response and high sensitivity along with long term stability. Recently, an ultrasensitive humidity sensor have been achieved by Zhang et al. [51] by incorporating polydopamine with graphene oxide and using the principles of quartz crystal microbalance. However, GO based sensors show changes due to proton conductivity, hence their selectivity may be distributed over gases donating or accepting protons. Therefore, selectivity is an area of concern and improvement.

5.3 Reduced graphene oxide-based humidity sensor

The reduction of insulating GO leads to formation of conductive rGO (reduced graphene oxide). The reduced rGO is sensitive to humidity owing to several defects and oxygen containing functional groups in the rGO surface. Incomplete reduction of GO leads to some capacitive behavior of the rGO based sensor. Guo et al. developed graphene oxide over PET substrate by simultaneous reduction and patterning with the help of two-beam-laser interference method [52]. The sensor fabricated showed quick response and recovery with very high sensitivity due to enhanced adsorption of water molecules facilitated by laser induced graphene surface. GO film fabricated through layer-by-layer covalent anchoring showed high sensitivity in the range 30–90% humidity with minimum hysteresis. The sensor exhibited response time of 28 s and recovery of 48 s with long duration stability [53]. Rapid thermal annealing was carried out on GO by Phan et al. to study the influence of oxygen functional groups on humidity sensing performance [54]. It gives an idea of the sensing properties varying with reduction degree of GO. The work showed that with increase in annealing temperature, GO exhibited reduction in resistivity and led to loss of its capability towards adsorbing water molecules, which ultimately led to reduction in the response of the sensor. A recent study by Shojaee et al. showed that there is variation in sensing performance with degree of reduction of the GO film. They reduced GO by hydrothermal method and the reaction time accounted for the degree of reduction [55]. They found that moderate reduction of GO accounted for optimized sensitivity with response and recovery time. This is because restoration of the sp2 carbon network contributes to the response time whereas the residual oxygen groups lead to the sensitivity of the GO film. A study by Papazoglou et al. elaborated on the reduction of GO by laser reduction and in-situ sequential laser transfer. The sensor exhibited a response time of less than one minute in water concentration 1700–20000 ppm having 1700 ppm as limit of detection [56].

Humidity sensing being a significant part of our daily life, rGO based flexible and wearable sensors have attracted much attention. Preparation of rGO based sensors in fiber substrate is one of the many strategies to make wearable sensor devices. Qu's group fabricated microfibers based on double helix core sheath rGO which was sensitive to multiple stimulus. Small changes in temperature, mechanical properties and RH led to significant current response [57]. A unique humidity sensor was fabricated by Choi et al. where the sensing layer was nitrogen doped rGO fibers with platinum nanoparticles deposited on the rGO surface. These nanoparticles behaved as dissociation catalysts while sensing humidity. A wide range of humidity (6.1–66.4%) was detected by the rGO fibers with 136% sensitivity. Wearable devices based on natural fibers such as silk or spider silk are employed to achieve biodegradability, superior mechanical properties and excellent skin affinity. A flexible humidity sensor was fabricated by Li et al. where silk fabrics were coated with nickel and GO sheets. The sensor exhibited very fast detection towards humidity and showed probable use in monitoring human respiration [58]. Silk interlayers were introduced into GO films by Ma et al. who proposed a very light weight and tough printable bio papers for humidity sensing applications [59].

5.4 Graphene/2D materials-based humidity sensor

Recently, ultrasensitive sensors based on 2D materials such as TMDCs (transition metal dichalcogenides) like WS2 and MoS2 and black phosphorus have gained significant attention owing to their novel electronic properties and structures. Graphene has been incorporated with WS2, MoS2 and black phosphorus as nanocomposites for enhanced performance as humidity sensors. The first humidity sensor based on MoS2/GO was developed by Burman et al. which exhibited extremely high response of 1600 times at 85% RH. This high response was due to enhanced proton conductivity in MoS2 and GO [60]. Another work based on MoS2/rGO composites were reported by Park et al., where they prepared the composites without any additional heating or additives. Then the composite material was drop casted over interdigitated electrodes for sensing humidity [61]. The MoS2/rGO composites exhibited 200 times enhanced response towards water molecules when compared to pure rGO based sensor. Formation of a p-n heterojunction between rGO and MoS2 led to this remarkable improvement in the sensing performance. Hydrothermal method to fabricate rGO/MoS2 composites for high performance humidity sensor was developed by Park et al. [62]. Jha et al. developed GO/WS2 composite for sensing humidity exhibiting 65.8 times response at 40% RH and 590 times response at 80% RH. The sensor exhibited a response time of 25 s and a recovery time of 29 s which was due to improved proton conductivity at the GO/WS2 interface due to oxygen linking activities [63]. The 2D layer and crystalline nature of black phosphorus induced water molecules to naturally adsorb onto the surface attributing to ultra-sensitive nature of black phosphorus. However, it suffered from repeatability due to instability of water molecules with black phosphorus. This limitation was overcome by Phan et al. when they developed black phosphorus/graphene composite by introducing graphene into black phosphorus. The stability of the senor was taken care by the interface formed between graphene and black phosphorus. The sensor showed linearity in response within 15–70% RH with a response of 43.4% at 70% RH. Response time of 9 s and recovery time of 30 s was noted for this composite structure. Humidity sensors based on graphene/2D material shows potential, among which MoS2/rGO composites have maximum potential. However, the response and recovery time of the sensors must be reduced for achieving optimum performance.

5.5 Graphene quantum dots as humidity sensor

Graphene quantum dots (GQDs) exhibits almost similar properties and structure as layered-graphene, however its electronic properties are based upon electronic states at the edge and size of the quantum dots (exhibiting quantum confinement). The size can be controlled to manipulate the desired properties of GQD. Sreeprasad et al. developed a humidity sensor based on percolating network of GQDs which was assembled selectively over a polyelectrolyte microfiber. The work explained electron-tunneling modulation in the network of GQDs. Here GQDs was responsible for electron transportation and the polymer was responsible for water mass transfer [64]. The presence of water vapor reduced the width of tunneling barrier between GQDs by 0.36 nm and enhanced its conductivity 43 times. Ruiz et al. developed a resistive humidity sensor based on GQDs which was prepared by pyrolysis of citric acid over interdigitated electrodes [65]. An exponential variation of sensitivity over the RH range 15–80% was observed along with a quick response of ~5 s. The water molecules condensed over the GQD surface through capillary action which accounted for enhanced sensing performance. A similar approach was taken by Alizadeh et al. to prepare GQD based humidity sensor [66], which showed enormous sensitivity to humidity variation in the environment with response time close to 10 s. Interestingly they reported two separate sensing phenomena for 0–52% RH and 52–97% RH. When the RH range was low, the electrical resistance decreased as the adsorbed water molecules injected hole carriers. However, at the high RH range, the ionic proton transportation improved due to the adsorption of water molecules and caused a reduction in electrical resistance of the sensor. For the first time a flexible humidity sensor based on GQD was demonstrated by Hosseini et al. where a facile hydrothermal method was used to synthesize GQDs. The synthesized GQDs were drop casted on interdigitated electrodes fabricated over a flexible polyimide substrate (**Figure 3(a)**) [66]. The GQD film exhibited porous like structure which enhanced the sensing performance as shown in **Figure 3(b)**. The fabricated sensor exhibited exponential characteristics in response towards 12–100% humidity range. The response and recovery time of the sensor was found to be 12 s and 43 s respectively as depicted in **Figure 3(c)**. The sensor showed a quick response when exposed to human breath flow (90% RH) as shown in **Figure 3(d)**. Hence, it is found that GQDs have certain benefits as humidity sensor such as good selectivity, enhanced response and very fast response

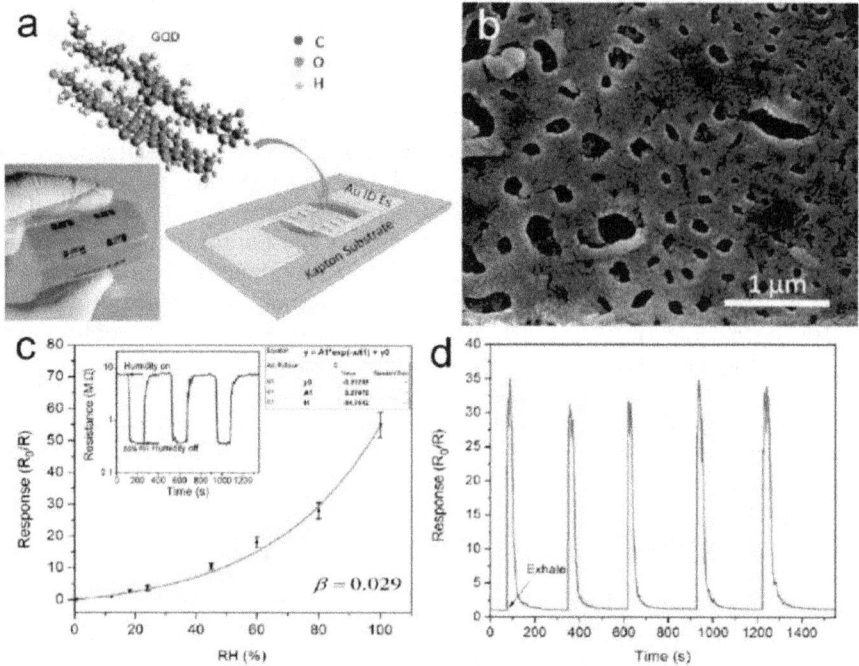

Figure 3.
(a) Schematic of flexible sensor; (b) FESEM image of the sensing material; (c) response vs. RH plot; (d) response to human breath. Reproduced with permission from [66], copyright the Royal Society of Chemistry, 2017.

and recovery. Few more studies on GQDs (especially on stability) can establish it on commercial platform as potential humidity sensor.

5.6 Graphene/metal or graphene/metal oxide as humidity sensor

Humidity sensing performance of metal oxides such as ZnO, SnO2, CuO was observed due to diversity in morphology, high surface to volume ratio, presence of defects and vacancies. However, limitation was observed in conductivity and they exhibited slow electron diffusion which was responsible for reduced response. On the other hand, GO or rGO exhibited lack of reversibility. Therefore, a graphene/metal oxide composite may enhance the sensing performance optimally.

Humidity sensors based on graphene/ SnO2 composite has been widely studied, where the humidity sensing performance was observed to improve due to incorporation of graphene over SnO2. A study on SnOx coated with graphene on carbon fibers (graphene/SnOx/CF) [67] exhibited sensitivity of 6.22 which is 2 times higher than the uncoated one [68]. A humidity sensor based on SnO2/graphene wrapped with GO was developed by Xu et al. which showed very fast response and recovery (less than 1 s) and enormous sensitivity of 32 MΩ/RH%. This high sensitivity and good stability of the sensor was due to enhanced conductivity of graphene and oxygen rich functional groups of GO (specially hydroxyl and epoxy). A one step facile hydrothermal method was used to fabricate rGO/SnO2 composite by Zhang et al. The composite material after fabrication was drop casted on microelectrodes for developing the humidity sensor. The sensor exhibited enhanced sensitivity, fast response and recovery when compared to pristine rGO. The improvement in sensitivity was dedicated to the defects and vacancies introduced by SnO2

Sl. No	Sensing layer	Working principle/Measurement parameters	Reference
1.	Graphene oxide	Capacitive	[32]
2.	Graphene oxide thin film	Surface acoustic wave atomizer	[33]
3.	Graphene oxide	Optical fiber	[39]
4.	Graphene	Field effect transistor	[27]
5.	SnS2-Reduced graphene oxide	Resistive	[29]
6.	Graphene	Resistive	[42]
7.	Polydopamine/ graphene oxide	Quartz crystal microbalance	[51]
8.	Graphene quantum dots	Resistive	[65]
9.	Reduced graphene oxide/CuO	Resistive	[70]
10.	Self-powered Graphene oxide	Open circuit voltage and current	[74]

Table 1.
Shows the evolution of graphene-based humidity sensors based upon the sensing layer and the different working principles.

nanoparticles and the interface formed between the two materials. Investigation of the same material as humidity sensor was also carried in resistive mode [69]. Fe doped SnO_2 nanoparticles when employed with rGO sheets displayed highly improved sensing performance [69]. A rGO/CuO nano-composite was synthesized by Wang et al. which showed relatively good sensing characteristics [70]. The sensor showed fast response and high sensitivity which was mainly due the formation of Schottky junction between the two materials. It has been demonstrated that even incorporation of graphene on ZnO or TiO_2 nanoparticles enhance the performance of the humidity sensors [71].

There are few reports on humidity sensors based on composite of graphene and metal nanoparticles. A molecular combing method to prepare GO-Ag scrolls was developed by Liu et al. where the Ag nanoparticles were deposited uniformly over the GO surface [72]. The material when reduced by hydrazine to rGO/Ag scrolls showed a response of 3 orders in magnitude compared to that of bare rGO scrolls when exposed to humid atmosphere. The improvement in conductivity of the rGO/Ag scrolls due to encapsulation of the silver nanoparticles contributed towards excellent sensitivity of the sensor. Conduction pathways formed by Ag nanoparticles on the surface of rGO by self-assembly led to improvement of response and linearity of the sensor device. An interesting work by Yeo et al. showed the suppression of humidity dependence of the sensor based on rGO due to incorporation of Cu nanoparticles. The Cu nanoparticles were used to reduce the electrical resistance to detect various other gases (**Table 1**) [73].

6. Future outlook and summary

Different mechanisms have been attempted to develop highly efficient humidity sensors based on graphene. Pristine graphene-based humidity sensors exhibit high response with 1 ppm detection limit; however, it faces issues such as poor recovery and limited selectivity. The recovery and selectivity limitations can be overcome by chemical modification of the graphene surface with desired functional groups. Humidity sensors based on graphene oxide shows good sensitivity and quick response in the higher range of humidity. However, at low humidity conditions (less than 5%), detection is challenging. At higher humidity levels, these graphene oxide sensors exhibit relatively poor stability and swelling effect. Complicated circuits are required for impedance or capacitive working mode. Reduced graphene oxide, another counterpart of graphene facilitates development of resistive humidity sensors with very easy methods of fabrication. These sensors consume very less power and are easy to detect. Reduced graphene oxide has lesser number of oxygen functional groups compared to graphene oxide, which leads to reduction in sensing response. Different other approaches such as incorporating graphene with metal nanoparticles, metal oxide nanoparticles or 2D layered materials have enormously improved the sensitivity, range of detection, negligeable hysteresis and quick response. However, these devices suffer from long duration stability and reproducibility.

High performance humidity sensors based on graphene is still required with sufficiently high sensitivity, wide range of humidity, high degree of selectivity, quick recovery and response time and negligeable hysteresis. The foremost important requirement of the graphene-based sensor is long term stability, repeatability and full recovery. The degree of oxidation and reduction of graphene sheets possess a challenge towards reproducibility of the graphene- based sensors. The formation of graphene oxide and its reduction needs precise control, failing to which leads to different sensing performance. Improvement is also required in terms of detection range of humidity. Sensing at low humidity is also an important parameter. At last, it is very important now to consider issues related to commercialization of graphene- based humidity sensors. The points to be considered are repeatability, long term stability, large scale integration, packaging and anti-chemical characteristics. An industrially manufactured digital biosensor based on graphene is developed by Goldsmith et al. [75], however it has a long way to go towards commercialization. Most of the studies in humidity sensor based on graphene have not considered the issue of power consumption, which is one of the key focus areas while developing wearable electronics.

Acknowledgements

The authors would like to thank Microelectronics and MEMS laboratory, IIT Kharagpur for supporting this work.

Author details

Avik Sett[1], Kunal Biswas[2], Santanab Majumder[3], Arkaprava Datta[3]
and Tarun Kanti Bhattacharyya[1*]

1 Department of Electronics and Electrical Communication Engineering,
IIT Kharagpur, Kharagpur, West Bengal, India

2 Department of Biotechnology, Maulana Abul Kalam Azad University of
Technology, West Bengal, India

3 School of Nanoscience and Technology, IIT Kharagpur, Kharagpur, West Bengal,
India

*Address all correspondence to: tkb@ece.iitkgp.ac.in

IntechOpen

References

[1] Seiyama, T., Yamazoe, N. and Arai, H., 1983. Ceramic humidity sensors. Sensors and Actuators, 4, pp.85-96.

[2] Chen, Z. and Lu, C., 2005. Humidity sensors: A review of materials and mechanisms. Sensor letters, 3(4), pp.274-295.

[3] Blank, T.A., Eksperiandova, L.P. and Belikov, K.N., 2016. Recent trends of ceramic humidity sensors development: A review. Sensors and Actuators B: Chemical, 228, pp.416-442.

[4] Manikandan, V., Sikarwar, S., Yadav, B.C. and Mane, R.S., 2018. Fabrication of tin substituted nickel ferrite (Sn-NiFe2O4) thin film and its application as opto-electronic humidity sensor. Sensors and Actuators A: Physical, 272, pp.267-273.

[5] Tripathy, A., Pramanik, S., Manna, A., Bhuyan, S., Azrin Shah, N.F., Radzi, Z. and Abu Osman, N.A., 2016. Design and development for capacitive humidity sensor applications of lead-free Ca, Mg, Fe, Ti-oxides-based electro-ceramics with improved sensing properties via physisorption. Sensors, 16(7), p.1135.

[6] Li, W., Liu, J., Ding, C., Bai, G., Xu, J., Ren, Q. and Li, J., 2017. Fabrication of ordered SnO2 nanostructures with enhanced humidity sensing performance. Sensors, 17(10), p.2392.

[7] A. Sett, A.K.Mukhopadhyay, Monojit Mondal, Santanab Majumder and T.K.Bhattacharyya, 2018. Tuning surface defects of mesoporous ZnO nanorods for high-speed humidity sensing applications. IEEE Sensors Conference 2018.

[8] Morais, R.M., dos Santos Klem, M., Nogueira, G.L., Gomes, T.C. and Alves, N., 2018. Low cost humidity sensor based on PANI/PEDOT: PSS printed on

paper. IEEE Sensors Journal, 18(7), pp.2647-2651.

[9] Zhao, J., Li, N., Yu, H., Wei, Z., Liao, M., Chen, P., Wang, S., Shi, D., Sun, Q. and Zhang, G., 2017. Highly sensitive MoS2 humidity sensors array for noncontact sensation. Advanced materials, 29(34), p.1702076.

[10] Guo, H., Lan, C., Zhou, Z., Sun, P., Wei, D. and Li, C., 2017. Transparent, flexible, and stretchable WS 2 based humidity sensors for electronic skin. Nanoscale, 9(19), pp.6246-6253.

[11] Cho, S.Y., Lee, Y., Koh, H.J., Jung, H., Kim, J.S., Yoo, H.W., Kim, J. and Jung, H.T., 2016. Superior chemical sensing performance of black phosphorus: Comparison with MoS2 and graphene. Advanced Materials, 28(32), pp.7020-7028.

[12] Zhou, G., Byun, J.H., Oh, Y., Jung, B.M., Cha, H.J., Seong, D.G., Um, M.K., Hyun, S. and Chou, T.W., 2017. Highly sensitive wearable textile-based humidity sensor made of high-strength, single-walled carbon nanotube/poly (vinyl alcohol) filaments. ACS applied materials & interfaces, 9(5), pp.4788-4797.

[13] Varghese, S.S., Lonkar, S., Singh, K.K., Swaminathan, S. and Abdala, A., 2015. Recent advances in graphene based gas sensors. Sensors and Actuators B: Chemical, 218, pp.160-183.

[14] Yavari, F. and Koratkar, N., 2012. Graphene-based chemical sensors. The journal of physical chemistry letters, 3(13), pp.1746-1753.

[15] Park, S., An, J., Suk, J.W. and Ruoff, R.S., 2010. Graphene-based actuators. Small, 6(2), pp.210-212.

[16] Lee, C., Wei, X., Kysar, J.W. and Hone, J., 2008. Measurement of the

elastic properties and intrinsic strength of monolayer graphene. Science, 321(5887), pp.385-388.

[17] Lipson, H.S. and Stokes, A.R., 1942. The structure of graphite. Proceedings of the Royal Society of London. Series A. Mathematical and Physical Sciences, 181(984), pp.101-105.

[18] Abbott's, I.E., 2007. Graphene: Exploring carbon flatland. Phys. Today, 60(8), p.35.

[19] Geim, A.K., 2009. Graphene: Status and prospects. Science, 324(5934), pp.1530-1534.

[20] Geim, A.K. and Novoselov, K.S., 2010. The rise of graphene. In nanoscience and technology: A collection of reviews from nature journals (pp. 11-19).

[21] Lee, C., Wei, X., Kysar, J.W. and Hone, J., 2008. Measurement of the elastic properties and intrinsic strength of monolayer graphene. Science, 321(5887), pp.385-388.

[22] Chen, J., Yao, B., Li, C. and Shi, G., 2013. An improved hummers method for eco-friendly synthesis of graphene oxide. Carbon, 64, pp.225-229.

[23] Chaste, J., Saadani, A., Jaffre, A., Madouri, A., Alvarez, J., Pierucci, D., Aziza, Z.B. and Ouerghi, A., 2017. Nanostructures in suspended mono-and bilayer epitaxial graphene. Carbon, 125, pp.162-167.

[24] Fogarassy, Z., Rümmeli, M.H., Gorantla, S., Bachmatiuk, A., Dobrik, G., Kamarás, K., Biró, L.P., Havancsák, K. and Lábár, J.L., 2014. Dominantly epitaxial growth of graphene on Ni (1 1 1) substrate. Applied surface science, 314, pp.490-499.

[25] Tetlow, H., De Boer, J.P., Ford, I.J., Vvedensky, D.D., Coraux, J. and Kantorovich, L., 2014. Growth of epitaxial graphene: Theory and experiment. Physics reports, 542(3), pp.195-295.

[26] Schedin, F., Geim, A.K., Morozov, S.V., Hill, E.W., Blake, P., Katsnelson, M.I. and Novoselov, K.S., 2007. Detection of individual gas molecules adsorbed on graphene. Nature materials, 6(9), pp.652-655.

[27] He, Q., Wu, S., Yin, Z. and Zhang, H., 2012. Graphene-based electronic sensors. Chemical Science, 3(6), pp.1764-1772.

[28] Meng, F.L., Guo, Z. and Huang, X.J., 2015. Graphene-based hybrids for chemiresistive gas sensors. TrAC Trends in Analytical Chemistry, 68, pp.37-47.

[29] Shafiei, M., Bradford, J., Khan, H., Piloto, C., Wlodarski, W., Li, Y. and Motta, N., 2018. Low-operating temperature NO2 gas sensors based on hybrid two-dimensional SnS2-reduced graphene oxide. Applied Surface Science, 462, pp.330-336.

[30] Piloto, C., Shafiei, M., Khan, H., Gupta, B., Tesfamichael, T. and Motta, N., 2018. Sensing performance of reduced graphene oxide-Fe doped WO3 hybrids to NO2 and humidity at room temperature. Applied Surface Science, 434, pp.126-133.

[31] Gao, W., Singh, N., Song, L., Liu, Z., Reddy, A.L.M., Ci, L., Vajtai, R., Zhang, Q., Wei, B. and Ajayan, P.M., 2011. Direct laser writing of micro-supercapacitors on hydrated graphite oxide films. Nature nanotechnology, 6(8), pp.496-500.

[32] Bi, H., Yin, K., Xie, X., Ji, J., Wan, S., Sun, L., Terrones, M. and Dresselhaus, M.S., 2013. Ultrahigh humidity sensitivity of graphene oxide. Scientific reports, 3(1), pp.1-7.

[33] Balashov, S.M. and Balachova, O.V., 2012. Filho, AP; Bazetto, MCQ; de

Almeida, MG surface acoustic wave humidity sensors based on graphene oxide thin films deposited with the surface acoustic wave atomizer. ECS Trans, 49, pp.445-450.

[34] Wohltjen, H., 1984. Mechanism of operation and design considerations for surface acoustic wave device vapour sensors. Sensors and Actuators, 5(4), pp.307-325.

[35] Xuan, W., He, M., Meng, N., He, X., Wang, W., Chen, J., Shi, T., Hasan, T., Xu, Z., Xu, Y. and Luo, J.K., 2014. Fast response and high sensitivity ZnO/glass surface acoustic wave humidity sensors using graphene oxide sensing layer. Scientific reports, 4(1), pp.1-9.

[36] Xuan, W., He, X., Chen, J., Wang, W., Wang, X., Xu, Y., Xu, Z., Fu, Y.Q. and Luo, J.K., 2015. High sensitivity flexible lamb-wave humidity sensors with a graphene oxide sensing layer. Nanoscale, 7(16), pp.7430-7436.

[37] Kuznetsova, I.E., Anisimkin, V.I., Kolesov, V.V., Kashin, V.V., Osipenko, V.A., Gubin, S.P., Tkachev, S.V., Verona, E., Sun, S. and Kuznetsova, A.S., 2018. Sezawa wave acoustic humidity sensor based on graphene oxide sensitive film with enhanced sensitivity. Sensors and Actuators B: Chemical, 272, pp.236-242.

[38] Xiao, Y., Zhang, J., Cai, X., Tan, S., Yu, J., Lu, H., Luo, Y., Liao, G., Li, S., Tang, J. and Chen, Z., 2014. Reduced graphene oxide for fiber-optic humidity sensing. Optics express, 22(25), pp.31555-31567.

[39] Wang, Y., Shen, C., Lou, W. and Shentu, F., 2016. Polarization-dependent humidity sensor based on an in-fiber Mach-Zehnder interferometer coated with graphene oxide. Sensors and Actuators B: Chemical, 234, pp.503-509.

[40] Wang, Ning, Wenhao Tian, Haosheng Zhang, Xiaodan Yu, Xiaolei Yin, Yonggang Du, and Dailin Li. "An easily fabricated high performance Fabry-Perot optical fiber humidity sensor filled with graphene quantum dots." Sensors 21, no. 3 (2021): 806.

[41] Owji, Erfan, Hossein Mokhtari, Fatemeh Ostovari, Behnam Darazereshki, and Nazanin Shakiba. "2D materials coated on etched optical fibers as humidity sensor." Scientific Reports 11, no. 1 (2021): 1-10.

[42] Yavari, F., Kritzinger, C., Gaire, C., Song, L., Gulapalli, H., Borca-Tasciuc, T., Ajayan, P.M. and Koratkar, N., 2010. Tunable bandgap in graphene by the controlled adsorption of water molecules. Small, 6(22), pp.2535-2538.

[43] Son, Y.J., Chun, K.Y., Kim, J.S., Lee, J.H. and Han, C.S., 2017. Effects of chemical and physical defects on the humidity sensitivity of graphene surface. Chemical Physics Letters, 689, pp.206-211.

[44] Fan, X., Elgammal, K., Smith, A.D., Östling, M., Delin, A., Lemme, M.C. and Niklaus, F., 2018. Humidity and CO2 gas sensing properties of double-layer graphene. Carbon, 127, pp.576-587.

[45] Zhen, Z., Li, Z., Zhao, X., Zhong, Y., Zhang, L., Chen, Q., Yang, T. and Zhu, H., 2018. Formation of uniform water microdroplets on wrinkled graphene for ultrafast humidity sensing. Small, 14(15), p.1703848

[46] Qadir, A., Sun, Y.W., Liu, W., Oppenheimer, P.G., Xu, Y., Humphreys, C.J. and Dunstan, D.J., 2019. Effect of humidity on the interlayer interaction of bilayer graphene. Physical review B, 99(4), p.045402.

[47] Yao, Y., Chen, X., Zhu, J., Zeng, B., Wu, Z. and Li, X., 2012. The effect of ambient humidity on the electrical properties of graphene oxide films. Nanoscale research letters, 7(1), pp.1-7.

[48] Park, E.U., Choi, B.I., Kim, J.C., Woo, S.B., Kim, Y.G., Choi, Y. and Lee, S.W., 2018. Correlation between the sensitivity and the hysteresis of humidity sensors based on graphene oxides. Sensors and Actuators B: Chemical, 258, pp.255-262.

[49] Chi, Hong, Lim Jun Ze, Xuemin Zhou, and Fuke Wang. "GO film on flexible substrate: An approach to wearable colorimetric humidity sensor." Dyes and Pigments 185 (2021): 108916

[50] Cai, J., Lv, C. and Watanabe, A., 2016. Laser direct writing of high-performance flexible all-solid-state carbon micro-supercapacitors for an on-chip self-powered photodetection system. Nano Energy, 30, pp.790-800.

[51] Zhang, Dongzhi, Xiaoshuang Song, Zhao Wang, and Haonan Chen. "Ultra-highly sensitive humidity sensing by polydopamine/graphene oxide nanostructure on quartz crystal microbalance." Applied Surface Science 538 (2021): 147816.

[52] Guo, L., Jiang, H.B., Shao, R.Q., Zhang, Y.L., Xie, S.Y., Wang, J.N., Li, X.B., Jiang, F., Chen, Q.D., Zhang, T. and Sun, H.B., 2012. Two-beam-laser interference mediated reduction, patterning and nanostructuring of graphene oxide for the production of a flexible humidity sensing device. Carbon, 50(4), pp.1667-1673.

[53] Su, P.G. and Chiou, C.F., 2014. Electrical and humidity-sensing properties of reduced graphene oxide thin film fabricated by layer-by-layer with covalent anchoring on flexible substrate. Sensors and Actuators B: Chemical, 200, pp.9-18.

[54] Phan, D.T. and Chung, G.S., 2015. Effects of rapid thermal annealing on humidity sensor based on graphene oxide thin films. Sensors and Actuators B: Chemical, 220, pp.1050-1055.

[55] Shojaee, M.H.S.M., Nasresfahani, S., Dordane, M.K. and Sheikhi, M.H., 2018. Fully integrated wearable humidity sensor based on hydrothermally synthesized partially reduced graphene oxide. Sensors and Actuators A: Physical, 279, pp.448-456.

[56] Papazoglou, S., Petridis, C., Kymakis, E., Kennou, S., Raptis, Y.S., Chatzandroulis, S. and Zergioti, I., 2018. In-situ sequential laser transfer and laser reduction of graphene oxide films. Applied physics letters, 112(18), p.183301.

[57] Zhao, F., Zhao, Y., Cheng, H. and Qu, L., 2015. A graphene fibriform responsor for sensing heat, humidity, and mechanical changes. Angewandte Chemie, 127(49), pp.15164-15168.

[58] Li, B., Xiao, G., Liu, F., Qiao, Y., Li, C.M. and Lu, Z., 2018. A flexible humidity sensor based on silk fabrics for human respiration monitoring. Journal of Materials Chemistry C, 6(16), pp.4549-4554.

[59] Ma, R. and Tsukruk, V.V., 2017. Seriography-guided reduction of graphene oxide biopapers for wearable sensory electronics. Advanced functional materials, 27(10), p.1604802.

[60] Burman, D., Ghosh, R., Santra, S. and Guha, P.K., 2016. Highly proton conducting MoS 2/graphene oxide nanocomposite based chemoresistive humidity sensor. Rsc Advances, 6(62), pp.57424-57433.

[61] Park, S.Y., Kim, Y.H., Lee, S.Y., Sohn, W., Lee, J.E., Shim, Y.S., Kwon, K.C., Choi, K.S., Yoo, H.J., Suh, J.M. and Ko, M., 2018. Highly selective and sensitive chemoresistive humidity sensors based on rGO/MoS 2 van der Waals composites. Journal of Materials Chemistry A, 6(12), pp.5016-5024.

[62] Park, S.Y., Lee, J.E., Kim, Y.H., Kim, J.J., Shim, Y.S., Kim, S.Y., Lee, M.H. and

Jang, H.W., 2018. Room temperature humidity sensors based on rGO/MoS2 hybrid composites synthesized by hydrothermal method. Sensors and Actuators B: Chemical, 258, pp.775-782.

[63] Jha, R.K., Burman, D., Santra, S. and Guha, P.K., 2017. WS 2/GO nanohybrids for enhanced relative humidity sensing at room temperature. IEEE Sensors Journal, 17(22), pp.7340-7347.

[64] Sreeprasad, T.S., Rodriguez, A.A., Colston, J., Graham, A., Shishkin, E., Pallem, V. and Berry, V., 2013. Electron-tunneling modulation in percolating network of graphene quantum dots: Fabrication, phenomenological understanding, and humidity/pressure sensing applications. Nano letters, 13(4), pp.1757-1763.

[65] Ruiz, V., Fernández, I., Carrasco, P., Cabañero, G., Grande, H.J. and Herrán, J., 2015. Graphene quantum dots as a novel sensing material for low-cost resistive and fast-response humidity sensors. Sensors and Actuators B: Chemical, 218, pp.73-77.

[66] Hosseini, Z.S., Ghiass, M.A., Fardindoost, S. and Hatamie, S., 2017. A new approach to flexible humidity sensors using graphene quantum dots. Journal of Materials Chemistry C, 5(35), pp.8966-8973.

[67] Xu, J., Gu, S. and Lu, B., 2015. Graphene and graphene oxide double decorated SnO 2 nanofibers with enhanced humidity sensing performance. RSC advances, 5(88), pp.72046-72050.

[68] Fu, T., Zhu, J., Zhuo, M., Guan, B., Li, J., Xu, Z. and Li, Q., 2014. Xu, J., Gu, S. and Lu, B., 2015. Graphene and graphene oxide double decorated SnO 2 nanofibers with enhanced humidity sensing performance. RSC advances, 5(88), pp.72046-72050.Journal of Materials Chemistry C, 2(24), pp.4861-4866.

[69] Zhang, D., Chang, H. and Liu, R., 2016. Humidity-sensing properties of one-step hydrothermally synthesized tin dioxide-decorated graphene nanocomposite on polyimide substrate. Journal of Electronic Materials, 45(8), pp.4275-4281.

[70] Toloman, D., Popa, A., Stan, M., Socaci, C., Biris, A.R., Katona, G., Tudorache, F., Petrila, I. and Iacomi, F., 2017. Reduced graphene oxide decorated with Fe doped SnO2 nanoparticles for humidity sensor. Applied Surface Science, 402, pp.410-417.

[71] Wang, Z., Xiao, Y., Cui, X., Cheng, P., Wang, B., Gao, Y., Li, X., Yang, T., Zhang, T. and Lu, G., 2014. Humidity-sensing properties of urchinlike CuO nanostructures modified by reduced graphene oxide. ACS applied materials & interfaces, 6(6), pp.3888-3895.

[72] Liu, Y., Wang, L., Zhang, H., Ran, F., Yang, P. and Li, H., 2017. Graphene oxide scroll meshes encapsulated Ag nanoparticles for humidity sensing. RSC advances, 7(64), pp.40119-40123.

[73] Yeo, C.S., Kim, H., Lim, T., Kim, H.J., Cho, S., Cho, K.R., Kim, Y.S., Shin, M.K., Yoo, J., Ju, S. and Park, S.Y., 2017. Copper-embedded reduced graphene oxide fibers for multi-sensors. Journal of Materials Chemistry C, 5(48), pp.12825-12832.

[74] Ejehi, Faezeh, Raheleh Mohammadpour, Elham Asadian, Pezhman Sasanpour, Somayeh Fardindoost, and Omid Akhavan. "Graphene oxide papers in nanogenerators for self-powered humidity sensing by finger tapping." Scientific reports 10, no. 1 (2020): 1-11.

[75] Goldsmith, B.R., Locascio, L., Gao, Y., Lerner, M., Walker, A., Lerner, J., Kyaw, J., Shue, A., Afsahi, S., Pan, D. and Nokes, J., 2019. Digital biosensing by foundry-fabricated graphene sensors. Scientific reports, 9(1), pp.1-10.

Chapter 6

A Research on Polyamide6.6/ Polyurethane Blends in Finishing Process Which Are Used for Sportswear

Meliha Oktav Bulut and Ayşen Cire

Abstract

In this work, softeners obtained from various companies were applied to the polyamide6.6/polyurethane fabrics which are used in sportswear industry using impregnation and exhaustion methods; water vapor permeability were determined for humidity control, air permeability and capillarity tests of these fabrics were studied. In addition, the wool hydrolysate obtained from the waste wool was also applied to these fabrics by using exhaustion method and the fabric properties were compared. In order to investigate the washing resistance of the process, experiments were performed with 1% and 3% potassium aluminum sulfate $KAl(SO_4)_2$ and aluminum sulfate $Al_2(SO_4)$ under the same conditions. The chemical and morphological surface properties of the fabrics were examined by using X-ray photoelectron spectroscopy (XPS). It was observed that the capillarity, water vapor and air permeability, and handle values of fabrics treated with wool hydrolysate were better and more resistant to consecutive washings than the fabrics treated with commercial recipes. Furthermore, this process did not have a side effect on the color difference and whiteness values of the fabrics treated with wool hydrolysate. Thus, an example of sustainable, economical and environmental study was done.

Keywords: polyamide6.6/polyurethane, capillarity, wool hydrolysate, water vapor permeability, handle, sustainable, sportswear industry

1. Introduction

Sports activities have become a hobby and lifestyle for many people, since the importance of healthy life is known and the quality of life has increased. In addition, sports activities have become a necessity for today's people who want to get away from excessive work and overwhelming business of urban life. For this reason, the interest in sportswear has also increased. Not only professional athletes but also individuals who do sports are accustomed to wearing sportswear making the clothing a functional necessity.

Sportswear is regarded as an area open to development, high potential and high added value in the textile industry. The global sports apparel market grew up to 181 billion U.S. dollars in 2019, and compared with the previous year, it has increased more than seven billion U.S. dollars. It is estimated that it will continue to grow and reach approximately 208 billion U.S. dollars in 2025 [1, 2]. With the development of technical textiles, it is possible to have comfort during the highly intense sports activity and under different climate conditions. Knowing the humidity (water vapor permeability) and air permeability values of the clothing are two most important parameters in the measurement of comfort parameters as well as touching. Therefore, the measurement of how humid is a textile material is of great importance as well. For such measurement of humidity not only traditional testing methods but also recently developed sensors with different techniques – such as inkjet printing, carbon nanotubes, coating technology, stamp transfer, electrospinning and dip coating-can be used [3, 4]. Humidity control is widely carried out in many sectors dominating the daily life, such as agriculture, chemistry, food, health, pharmacy and automation.

The hydrophilicity of natural fibers such as cotton provides superiority during sports activities. Since drying duration is too long, it reduces comfort and may cause various complications and discomforts [5, 6]. Otherwise, the traditional filaments such as polyester and polyamide are hydrophobic and are prone to dry rapidly and give a feeling of dryness. Synthetic yarn or blended yarn are used to increase the comfort property of the fabric [7, 8]. The water vapor and air permeability values of the garment depend on fiber, yarn species [9–11], surface structure [10–12], and finishing treatment [13–16]. A special finishing process is performed to increase hydrophilicity values of the fabric with synthetic fibers such as polyamide and polyester. Textile chemistry manufacturers produce different hydrophilicity enhancers and textile dyehouses use these products. Although these chemicals are suitable for eco-textile and environmental standards, they are produced with chemical materials and processes. Hence, they may give harm to the nature and the user during production and consumption. They are also expensive and have some drawbacks such as staining and yellowing/color change problems [16]. In this regard, the attempts for sustainable and clean production continue.

Raw wool, which is not efficient for textile production, is an important source for biopolymer. Recycling of this easily accessible protein source and the production of keratin are important sources for biocirculation and biocompatible material production. It has been used in cosmetics, recyclable composites, transportation, medical membranes, agriculture and coating industry in recent years. The wool obtained from sheep breed in Turkey is generally suitable for being used in products such as the blankets, rugs and carpet [17]. The gradual decrease in the carpet sector in Turkey has reduced the use of this wool and the material has only been waiting in warehouses [18]. The active wool can cause global warming due to methane gas which can be soluble in nature. As a result, the utilization/recovery of the wool waiting in the warehouse [19] has great importance, both in terms of obtaining materials which have superior properties and low cost, optimum utilization of resources and environmental protection [20]. The obtaining methods of keratin are reduction, oxidation, alkali and enzymatic hydrolysis [21–23]. Alkali, oxidation and reducing chemicals used break into disulfide and peptide bonds which are the basic structure of wool at high temperature and time, and wool solution is obtained. Studies using wool hydrolysate have been done to obtain fiber and nanofiber and to increase the performance of dyeing and starch material in textile industry.

The aim of this study is to obtain the hydrolysate from wool fibers, which were left in the carpet industry but now waiting in the warehouse as waste with the decrease in production, and the hydrolysate obtained was used instead of commercial finishing bath of the polyamide6.6/polyurethane fabrics which are used in sportswear industry in Turkey. Capillarity, water vapor and air permeability and handle and yellowing/color change values of the fabrics treated with wool hydrolysate were compared with the ones which were treated with commercial products [24]. Thus, a new production method has been introduced for the finishing of polyamide6.6/polyurethane blend, which are widely used in the market.

2. Experimental section

2.1 Fabrics

The fabrics made of 80/20 polyamide 6.6/polyurethane blend were used. The weights of both fabrics were 170 g/m^2 (150 den PA 6.6/30 den PU) and 190 g/m^2 (100 den PA6.6/30 den PU) and ready to dye. Fabric thicknesses were 0.48 mm and 0.66 mm respectively. The fabrics were knitted on a double comb bar (laying-in) Rachel warp knitting machine. While first laying-in bar (Gb1) were knitting the tuh pattern 1–0/2–3//with full draft, in the second bar (Gb2), tricot pattern 1–2/1–0// was being knitted with full draft.

2.2 Auxiliaries

- Arristan HPC T (hydrophility enhancer agent, polyester copolymer, non-ionic, CHT) [25].

- Tubingal SHE (Hydrophile silicone softener, functional polisiloxane, mild cationic, CHT/Bezama) [26].

- Hydroperm LPU liq c (Hidrophility enhancer agent, thermoreactive polyurethane resin, non-ionic, Archroma) [27].

- Siligen SIH liq (Hydrophile silicone softener, modified silicone, Archroma) [28].

- Potassium aluminum sulfate (KAl(SO$_4$)$_2$, Sigma-Aldrich)

- Aluminum sulfate (Al$_2$(SO$_4$) Merck)

- The hydrolysate wool solution: The solution was obtained by alkali hydrolysis by using wool fibers which have 28 micron fineness and 40–60 mm length [21]. Average particle size was detected 211.91 millimicron with Mastersize 2000 in Merlab/ODTU.

2.3 Methods

Recipe for impregnation method

- 50 g/lt Arristan HPC T or Hydroperm LPU liq

- 20 g/lt Tubingal SHE or Siligen SIH, pH 5–5.5 (CH$_3$COOH),

- The fabric was padded with impregnated liquid and then they were squeezed at pick up 70% and were dried in Mathis CH-8156, 110°C for 3 min.

Recipe for exhaustion method

- % 3.5 Arristan HPC T or Hydroperm LPU liq

- %1.4 Tubingal SHE or Siligen SIH liq, pH value: 5–5.5 (CH$_3$COOH) at liquid ratio of 10:1 in Ataç Lab-Dye HT 10 for 30 min, at 40°C.

Samples were treated by using both impregnation and exhaustion methods in the same conditions in which only wool solution was used instead of chemicals. In order to ensure the washing resistance of the process, additions of 1% and 3% KAl $(SO_4)_2$ and $Al_2(SO4)$ were worked under the same conditions.

Dye uptake experiments were carried out in an Atac LAB-DYE HT machine at liquor ratio of 10: 1 as shown in **Figure 1**.

Recipe 1

- 0.12% Nylosan Red N-2RBL (CI Acid Red 336)

- 0.14% Nylosan Blue N-BLN (CI Acid Blue 350)

- 0.55% Optilan Golden Yelow MF-RL (CI Acid Orange 67)

pH value: 5–5.5 (CH_3COOH).
Recipe 2

- 0.17% Nylosan Red N-2RBL (CI Acid Red 336)

- 0.006% Nylosan Blue N-BLN (CI Acid Blue 350)

- 0.50% Optilan Golden Yelow MF-RL (CI Acid Orange 67)

pH value: 5–5.5 (CH_3COOH).

After dyeing, the samples were bathed at 50°C for 10 minutes with a non-ionic detergent (Fluidol W 100, Pulcra Chemicals) and rinsed with cold water.

2.4 Measurements and characterizations

All the physical measurements following the process were carried out after conditioning the fabrics for 24 hours under the standard atmosphere conditions

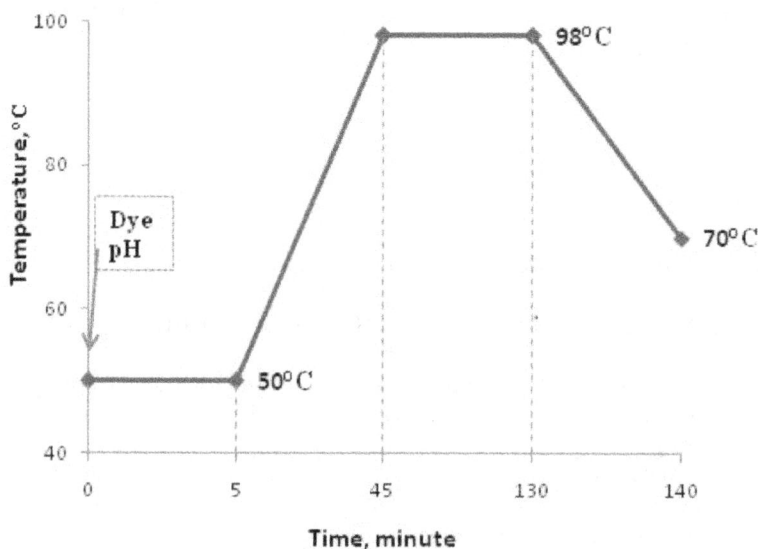

Figure 1.
Dyeing graph.

(20°C ± 2) temperature, % 65 ± 2). The capillarity of the fabrics was evaluated with the capillarity test method according to DIN 53924, water vapor permeability was measured according to ASTM E96-B and air permeability of the fabrics was measured by the Textest FX 3300 model Air Permeability Tester according to ASTM D737–04.

The laundry was done by using a front-loading Wascator machine (Electrolux FOM) with 2.0 kg loads consisting of processed samples and 100% PES ballast fabrics. All the washing cycles were performed according to BS EN ISO 26330 Standard (5A program). This laundering process was repeated 5 times in accordance with supplier's recommendation (Archroma and CHT). The samples were dried with dry flat in laboratory condition for 24 hours. The chemical and morphological surface properties of the fabrics were examined by using X-ray photoelectron spectroscopy (XPS). The samples were determined in terms of surface smoothness with XPS K-Alpha Surface Analysis with monochromatic Al Kα irradiation. The relative amounts of various bound atoms were determined through C1s, O 1 s, N1s, Si2p, Ca2p, S2p. Working condition is shown in **Table 1**.

The handle of the samples were carried out according to two different methods under the standard atmospheric conditions (20 °C ± 2) temperature, % 65 ± 2). Ten healthy women were selected as the participant group aged between 35 and 65 consisting of professionals including academic lecturers at Department of Textile Engineering, Suleyman Demirel University, academic lecturers at Textile, Apparel, Footwear and Leather Department of Technical High School at Isparta University of Applied Sciences and Dyehouse Manager of Isparta Mensucat Corporation.

Samples were examined on 10 subjects with 3 repetition in terms of thinnes/ thickness, softness/stiffness, smoothness/roughness and total handle values [29]. Moreover, another subjective evaluation was to have carried out examination in terms of softness/coolness/dampness sensation. The scale used is as shown in **Table 2**.

The degree of whiteness and yellowness indices of the samples were assessed by the CIE value and ASTM E 313 respectively, using Macbetch Coloreye 7000A. Color differences were indicated as ΔE, which was computed by Eq. (1):

$$\Delta E = \left[(\Delta L)^2 + (\Delta a)^2 + (\Delta b)^2 \right]^{1/2} \tag{1}$$

In the CIELAB color space, L is the lightness; a is the red/green axis, b is the yellow/blue axis, c is the chroma and h is the hue, ΔE is the color difference between the reference and the sample.

Parameter	
Total acquisition time	3 mins 24.2 secs
Number of scans	15
Source gun type	Al K Alpha
Spot size	400 μm
Lens mode	Standard
Analyzer mode	CAE: Pass Energy 150.0 eV
Energy step size	1.000 eV
Number of energy steps	1361

Table 1.
Working conditions used in XPS analysis.

Attribute	Rating scale			Time (s)
Thinness/thickness	1 **Thinnest**	5 **medium**	10 **thickest**	15
Softness/stiffness	1 softest	5 medium	10 stiffest	20
Smoothness/roughness	1 smoothest	5 medium	10 roughest	15
Total handle value	1 Not proper	5 medium	10 Most proper	20
Softness/coolness/dampness	1 Not proper	3 medium	5 Most proper	60

Table 2.
The scale used for subjective handle.

3. Results and discussion

3.1 The finishing process using impregnation and exhaustion methods

In **Table 3** the obtained capillarity, air permeability and water vapor permeability values of the sample values are given. These values are compared with those of fabrics non-treated.

As given in **Table 3**, the capillarity, air permeability and water vapor permeability values of the samples having two different weights are similar to each other when treated with both softener combinations by using impregnation method.

As seen in the values from **Table 3**, the capillarity values of both fabrics increase significantly after finishing process. Capillarity is essentially stated by the surface energy of the structure. In a textile structure, the surface energy is largely determined by the chemical structure of the exposed surface of the fiber. Hydrophilic fibers have a high surface energy; therefore, these fiber take up humidity quickly than hydrophobic fiber. Hydrophobic fibers conversely possess low surface energy and resist to humidity. Hydrophilic finishing can be used as enhancer in surface energy between face and back of the fabric to improve its ability to wick [30]. The greatest increase is observed in fabrics treated with only hydrophilicity enhancer for both fabrics. However, the lowest capillarity, water vapor and air permeability values are obtained with silicone softener for both fabrics. It is known that the capillarity of the fabric decreases by the treatment with amino silicon due to its hydrophobic character of silicone softener. The pores are also covered by the placement of the silicone on the fabric. This process also causes to reduction in water vapor and air permeability values of them. Recently, hydrophilic effective softening agents can be produced by modifying the fatty acid long chain in the silicone structure. Therefore, the decrease in **Table 3** values is not at an excessive amount. According to the results of **Table 3**, the water vapor and air permeability values of the fabrics are improved together with the capillarity values. The process of humidity transport in hydrophobic textile material take place in wicking, spreading and evaporation [31, 32]. The fabric evaporation and water vapor values become better as the rate of wicking increases.

Capillarity values obtained by applying wool hydrolysate to both fabrics by using impregnation method and then drying were obtained under the same conditions, and since the values were quite low, they are not given in **Table 3**. This indicates that the hydrolysate cannot be attached to the polyamide6.6/polyurethane blend fabrics. The capillarity, air permeability and water vapor permeability values

		Capillarity (sec)			Air permeability (l/m²/s)	Water vapor permeability (g/24 s/m²)
		10	30	60		
Fabric without treatment	Fabric I	11	25	39	870	521
	Fabric II	14	33	49	820	507
Arristan HPC T/Tubingal SHE	Fabric I	27	47	67	970	634
	Fabric II	33	52	66	920	628
Arristan HPC T	Fabric I	34	54	68	945	643
	Fabric II	24	44	61	900	631
Tubingal SHE	Fabric I	33	52	65	890	633
	Fabric II	23	50	63	920	623
Hydroperm LPU liq c/Siligen SIH liq	Fabric I	32	51	62	1010	650
	Fabric II	31	51	60	990	640
Hydroperm LPU liq c	Fabric I	35	52	62	985	655
	Fabric II	34	53	62	950	648
Siligen SIH liq	Fabric I	20	42	61	930	630
	Fabric II	18	34	54	905	620

Table 3.
Capillarity, water vapor and air permeability values of samples by using impregnation method.

of the samples with two different weights by using exhaustion method are shown in
Table 4.

When the **Tables 3** and **4** are compared, the capillarity test results of the
samples processed with Hydroperm LPU liq c/Siligen SIH liq in both fabric 1 and
fabric 2 using exhaustion method are higher. This can be attributed to the harmony
of the ionic character due to the fact that both chemicals are non-ionic.

An important point in the **Tables 3** and **4** is that the fabrics which have same
knitting structures (fabric 1 and 2) but with different weights and fineness of the
yarn are different. While fabric 1 was 170 g/m² (150 den PA6.6/30 den PU), fabric 2
190 g/m² was (100 den PA6.6/30 den PU). This shows that capillarity depends on
the diameter of the yarn forming the surface. In textile structures, the spaces
between fibers effectively form capillarities. Therefore, the narrower are the spaces
between these fibers, the greater is the ability of the textile to absorb moisture. The
construction of fabric that forms narrow capillarity has vital importance to pick up
moisture quickly [11, 12, 33].

As stated in **Table 4**, only wool hydrolysate was applied to fabrics as a softener
bath at 40°C and 50°C. The capillarity, water vapor and air permeability values
obtained are observed to increase with the application at 50°C. Raising the temperature from 40–50°C improves the fixation of the wool hydrolysate to the fabric. The
softening process of polyamide6.6 fabric is carried out at 40–50°C. Since glass
transition temperature of polyamide6.6 is 60–80°C [34, 35], processing at 60°C and
above may cause the previously dyed fabric to flow into the softening bath.

In order to increase the fixation of wool hydrolysate to the polyamide6.6/poly-
urethane fabric, aluminum sulfate, potassium aluminum sulfate which are used as
mordant in wool dyeing were added in an amount of 1% and 3% in the finishing
bath at 50°C. According to the **Table 4** results, potassium aluminum sulfate at the
amount of 1% gives the highest results for capillarity. This can be attributed to the

			Capillarity (sec)			Air permeability ($l/m^2/s$)	Water vapor permeability ($g/24 s/m^2$)
			10	30	60		
Fabric without treatment		Fabric I	11	25	39	850	521
		Fabric II	14	33	49	820	507
Arristan HPC T/Tubingal SHE	40°C	Fabric I	25	38	50	1100	630
		Fabric II	21	38	45	1050	630
	50 °C	Fabric I	36	49	61	1100	643
		Fabric II	32	48	57	1020	625
Arristan HPC T	50 °C	Fabric I	33	43	65	1050	643
		Fabric II	28	53	59	975	625
Tubingal SHE	50 °C	Fabric I	28	45	48	990	620
		Fabric II	22	35	35	900	616
Hydroperm LPU liq c/Siligen SIH liq	40°C	Fabric I	21	46	58	1100	640
		Fabric II	20	34	52	1060	632
	50 °C	Fabric I	35	46	64	1120	651
		Fabric II	21	39	55	1105	632
Hydroperm LPU liq c	50 °C	Fabric I	41	60	72	1050	655
		Fabric II	28	51	66	1000	648
Siligen SIH liq	50 °C	Fabric I	16	31	45	990	635
		Fabric II	13	23	30	980	626
Wool hydrolyzate	40°C	Fabric I	16	32	46	950	551
		Fabric II	15	27	35	925	550
	50 °C	Fabric I	17	30	39	980	560
		Fabric II	13	25	33	955	552
$Al_2(SO_4)$/wool hydrolyzate	1%	Fabric I	13	28	55	1085	642
		Fabric II	19	39	53	1015	640
	3%	Fabric I	30	50	67	1120	650
		Fabric II	35	50	65	1100	653
$KAl(SO_4)_2$/wool hydrolyzate	1%	Fabric I	25	50	70	1175	671
		Fabric II	34	55	70	1150	670
	3%	Fabric I	34	53	70	1200	673
		Fabric II	37	53	69	1150	677

Table 4.
Capillarity, water vapor and air permeability values of samples by using exhaustion method.

fact that potassium aluminum sulfate is pure and has high water solubility [36], molecular weight and high chelating property. Hence, it can bond wool hydrolysate to polyamide6.6 surface [37]. The fabrics processed have given approximately the same capillarity results shown in **Table 4**. This can be attributed to richness of wool hydrolysate rich in hydrophilic groups [38, 39] and better bonds to polyamide6.6 fabrics.

			Capillarity (sn)			Air permeability $(1/m^2/s)$	Water vapor permeability $(g/24\ s/m^2)$
			10	30	60		
Fabric without treatment		Fabric I	11	25	39	850	521
		Fabric II	14	33	49	820	507
Hydroperm LPU liq c/Siligen SIH liq	50 °C	Fabric I	16	26	35	970	555
		Fabric II	7	13	19	950	533
KAl(SO$_4$)$_2$/Wool Hydrolyzate	50 °C	Fabric I	31	46	60	1070	640
		Fabric II	16	26	42	1040	632

Table 5.
Capillarity, water vapor and air permeability values of samples by using exhaustion method after 5 consecutive washing.

The samples were compared in respect of capillarity, water vapor and air permeability of samples treated with the wool hydrolysate containing 1% potassium aluminum sulphate and with commercial recipe (Hydroperm LPU liq c and Siligen SIH liq) after 5 consecutive washing. As it is seen in **Table 5**, the values of the fabric treated with wool hydrolysate are rather higher than commercial chemicals in terms of capillarity, water vapor and air permeability values.

3.2 XPS analyses

XPS analysis of the 170 g/m2 fabrics one of which was applied commercial recipe and the other was applied wool hydrolysate containing 1% potassium aluminum sulphate before and after 5 consecutive washing steps are shown in **Figures 2–5** and **Table 6**.

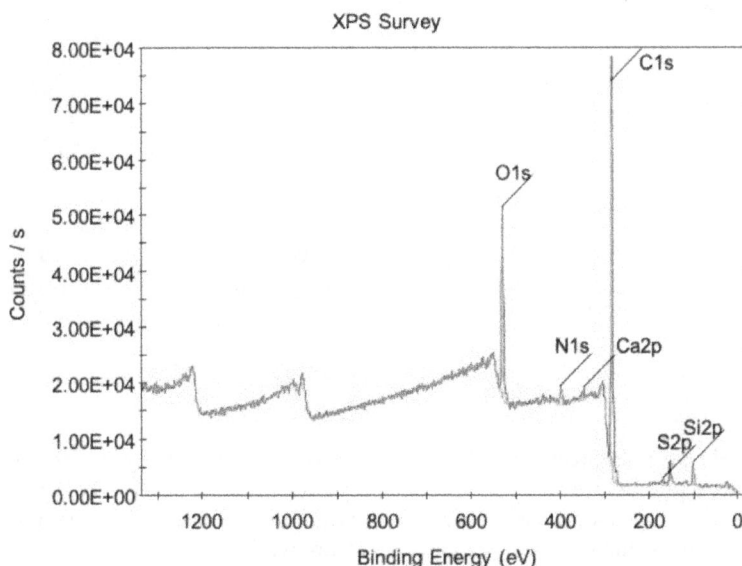

Figure 2.
The XPS analysis of fabric treated with the wool hydrolysate containing 1% KAl(SO$_4$)$_2$ before washing.

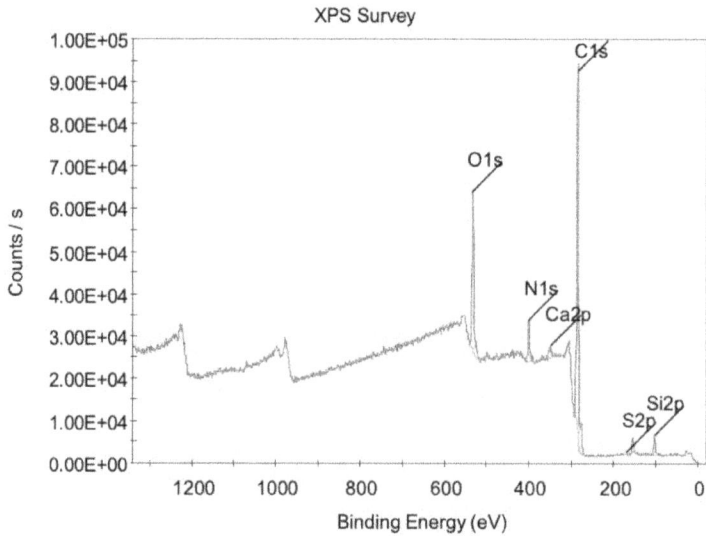

Figure 3.
The XPS analysis of wool hydrolysate after washing.

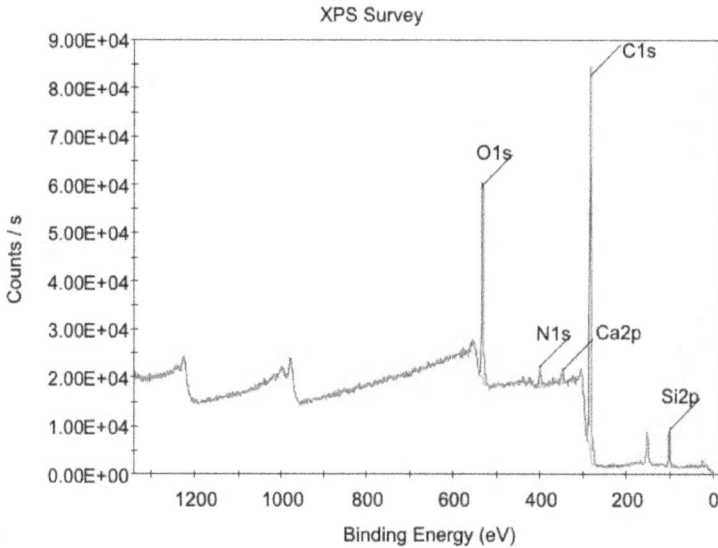

Figure 4.
The XPS analysis of the fabric treated with commercial recipe before washing.

In **Table 6** and **Figures 2** and **3**, XPS analyses of the fabric 170 g/m^2 treated with wool hydrolysate before and after 5 consecutive washing are observed. The carbon, oxygen and nitrogen atoms which are seen at the **Table 5** belong to polyamide6.6 structure. It is estimated that calcium comes from washing water bonded with amide onto the polyamide [40]. Although the process is made of demineralized water, it involves impurity and the amount of the calcium (Ca) increases after washing for all samples. Silicone (Si) is bonded to surface active agents based on dimethyl siloxane. This silicone comes from the silicone-based fats used in the

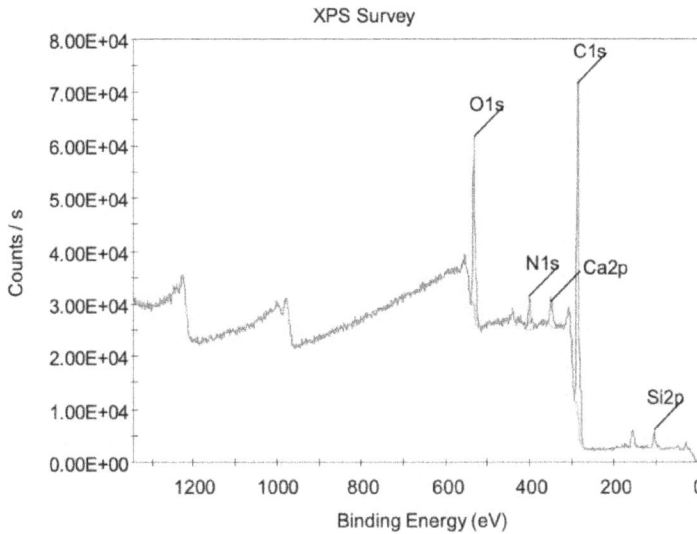

Figure 5.
The XPS analysis of the fabric treated with commercial recipe after washing.

polyurethane (PU) production [41, 42] and decreases with washing. Silicone has decreased from 8.96 to 7.74. The amount of sulfur (S) decreases with washing. It has fallen from %1.5 to %0.26 by getting away with wool hydrolysate.

Table 6 and **Figures 4** and **5** are the XPS analysis of the fabric 170 g/m^2 before and after washing according to the recipe at 50°C containing hyrophility enhancer and silicone (Hidroperm LPU liq.c and Siligen SIH liq). The difference from the analyses, they do not contain sulfur. Because this sulfur is situated in the structure of wool hydrolysate. This process was made of polyurethane resin and polysiloxane. Accordance with **Table 6** values, silicone has rised up at both two fabrics. This situation is the result of treating with micro silicone based polisiloxane. But the extremely decreasing of the amount of Si after 5 consecutive washing, indicates that the processing is not permanent and this explains that the capillarity and water vapor permeability values are lower than the samples treated with wool hydroly-sate. Silicone plays a significant role in the bounding of hydrophility enhancer agent to the fabric. Silicone can provide permanent effect with bounding hydroxyl group in fabric [32, 43].

3.3 The effect of the process on handle

One of the most important parameters to determine the effectiveness of a textile finishing is handle of fabric. The sensations of fabric such as softness, smoothness and drape created on the consumer can play a primary role in the preference of textiles.

So as to make subjective determination of fabric handles – as it is shown in measurement and characterization section, two different methods were used and the results are shown in **Tables 7** and **8**. The fabrics used have different weight, thickness and yarn count as it is mentioned in materials section. Furthermore, handle values of Fabric 1 were examined after 5 consecutive washing following 2 different softening process. As it is seen in **Table 7** values, while no increase were detected in thinness sensations of the fabrics treated with commercial finishing agents (Hyroperm LPU liq c/Siligen SIH liq), thickness sensation were detected in

The elemental properties of 170 g/m² fabric treated with wool hydrolyzate

Name	Peak BE	FWHM eV	Area (P) CPS.eV	Weight%	Q
C1s	284.96	3.36	290336.09	64.73	1
O1s	531.34	3.98	162004.01	19.89	1
Si2p	102.63	2.15	17248.82	8.96	1
N1s	399.61	4.05	17646.15	2.96	1
Ca2p	347.92	4.16	15059.93	1.96	1
S2p	168.90	3.03	5098.65	1.50	1

The elemental properties of the fabric weight 170 g/m² treated with wool
hydrolyzate after 5 consecutive washing

Name	Peak BE	FWHM eV	Area (P) CPS.eV	Weight%	Q
C1s	285.17	3.94	424943.11	68.37	1
O1s	531.31	4.24	194006.24	17.19	1
N1s	399.49	3.73	41320.20	4.99	1
Si2p	102.69	3.85	20666.95	7.74	1
Ca2p	347.58	6.36	15367.73	1.44	1
S2p	168.43	4.90	1207.10	0.26	1

The elemental properties of the fabric 170 g/m² treated with commercial
recipe

Name	Peak BE	FWHM eV	Area(P) CPS.eV	Weight %	Q
C1s	284.94	3.40	315966.69	61.93	1
O1s	531.74	3.73	184086.70	19.88	1
Si2p	102.21	3.34	29328.08	13.39	1
N1s	399.16	3.32	19843.50	2.92	1
Ca2p	347.29	3.27	16516.76	1.88	1

The elemental properties of the fabric 170 g/m² treated with commercial
recipe after 5 consecutive washing

Name	Peak BE	FWHM eV	Area (P) CPS.eV	Weight %	Q
C1s	285.80	5.51	422461.96	65.55	1
O1s	532.23	5.38	218497.26	18.68	1
N1s	400.16	5.34	39566.58	4.61	1
Si2p	103.29	5.25	19263.55	6.96	1
Ca2p	349.21	7.83	46517.72	4.21	1

Table 6.
XPS analyses of fabrics.

the fabrics treated with wool hydrolysate containing 1% potassium aluminum sulphate $(KAl(SO_4)_2$. This can be explained with the fact that the average particle size of the hydrolysate wool solution is much bigger than commercial finishing agents as it is mentioned auxialiries section. When the softness/stiffness and smoothness/roughness values were examined, it was determined that the fabrics treated with

	Fabric	Hyroperm LPU liq c/Siligen SIH liq	KAl(SO$_4$)$_2$/ wool hydrolysate	Hyroperm LPU liq c/Siligen SIH liq after 5 consecutive washing	KAl(SO$_4$)$_2$/wool hydrolysate after 5 consecutive washing
Thinness/ thickness	Fabric I	3.41	2.85	3.75	3.83	3.08
	Fabric II	4.58	4.75	5.23		
Softness/ stiffness	Fabric I	3.41	2.92	3.58	2.58	3.66
	Fabric II	4.33	4	2.16		
Smoothness/ roughness	Fabric I	3.41	3.17	4.25	2.91	3
	Fabric II	4.33	4.16	2.92		
Total handle value	Fabric I	5.66	8.58	7.41	7	7.16
	Fabric II	6.41	6.5	7.16		

Table 7.
Subjective evaluation of fabrics in terms of thinness/thickness, softness/stiffness, smoothness/roughness and total handle.

	Softness	Coolness	Dampness
Fabric I	3.5	3.41	3.16
Fabric II	2.66	3.66	4.08
Hyroperm LPU liq c/Siligen SIH liq			
Fabric I	4.21	3.42	3.5
Fabric I after 5 consecutive washing	4.08	3.75	3.83
Fabric II	3.66	3.25	3.33
KAl(SO$_4$)$_2$/wool hydrolysate			
Fabric I	4.17	4.25	4.08
Fabric I after 5 consecutive washing	4	4.25	4
Fabric II	4.08	4.42	4

Table 8.
Subjective evaluation of fabrics in terms of softness, coolness, dampness.

commercial finishing agents showed a bit higher sensation values. However, total handle values of fabrics processed with two different processes are similar as it is shown in **Table 7**.

In **Table 8**, handle values of samples were examined in terms of softness, coolness and dampness sensations. According to the **Table 8** values, all the fabrics treated with wool hydrolysate containing 1% potassium aluminum sulphate (KAl (SO$_4$)$_2$ show a considerable increase in coolness and dampness sensation values. This is true for all the fabrics processed with 5 consecutive washing. This is related to fabric's giving more coolness and dampnesss sensation by diffusing more damp-ness to the structure of the hydrolysate wool solution processed fabric with hydro-philic groups. This continues also after 5 consecutive washing stages.

According to the results above, the coolness and dampness sensations that the fabric gives depend on, to a great extent, the surface capillarity and water vapor permeability values. These results are in correlation with literature [44, 45]. This can be explained as the higher capillarity and water vapor transmission values cause humidity transfer which enables the fabric to have a greater evaporation capacity, and hence to have a more comfortable feeling.

3.4 The effect of the process on whiteness and color values of fabrics

Textiles are treated with a wide variety of complex chemicals in accordance with their end use. In addition to the production phase, softeners are the most well-known for their use in household and commercial cleaning. The basis of these softeners can be natural substances, such as modified animal fat, vegetable oil and wax, or hydrocarbon wax and silicon based synthetic materials. Due to the chemical nature of most softeners, they tend to turn yellow and change color with factors such as high temperature, prolonged storage, and their formulation [46]. In addition, due to its oily adhesive structure and application conditions (amount of use and pH), the increase in the amount taken causes the surface to turn yellow. The high free amine value of the cationic softener causes color change due to air oxidation during drying phase. The azo yellow and azoxy yellow resulting from the oxidation of the amino radical with the effect of heat and air cause the fabric to turn yellow [47]. Today, cationic softeners with ester quate structure that do not contain free amines can be preferred in colors not to cause yellowing.

The measurement of whiteness and yellowing index of the fabrics treated were done and shown in **Table 9**. According to the results of the **Table 9**, the values of the fabrics are slightly different.

In order to determine the color difference problem that softeners create in colored textile materials, the color difference value of the fabric treated with two different recipe were determined and given in **Tables 10** and **11**. According to the results of the **Tables 10** and **11**, the color difference of the fabrics treated with wool hydrolysate is similar.

Depending on the findings shown **Tables 9–11** yellowing/color change problems do not occur in the fabrics treated with commercial softener combination (Hydroperm LPU liq/Siligen SIH liq, thermoreactive polyurethane resin/modified hydrophile silicone softener) and wool hydrolysate prepared by diluting at a high rate (10 g/15 L).

	L	a	b	c	h°	WI-CIE/ Tint	YI- E313
Fabric without treatment	91.67	−0.86	−0.62	1.06	215.94	82.85/1.69	−1.93	1.69
	ΔL	Δa	Δb	Δc	Δh		YI- ASTM E313	ΔE
Wool hydrolyzate KAl (SO₄)₂ before washing	−0.58 D	0.27 R	−0.35 B	0.08 B	0.44 B	83.68/1.38	−2.44	0.66
Wool hydrolyzate KAl (SO₄)₂ after washing	−2.52 D	0.32 R	−0.19 B	−0.09 D	0.36 B	78.49/1.25	−2.11	1.02
Hydroperm LPU liq c/Siligen SIH liq before washing	−3.72 D	0.22 R	−0.10 B	−0.09 D	0.22 B	75.44/1.41	−2.03	1.34
Hydroperm LPU liq c/Siligen SIH liq after washing	−1.59 D	0.28 R	−0.19 B	−0.06 D	0.33 B	80.58/1.31	−2.13	0.74

Table 9.
Whiteness and yellowing index of the fabrics.

Fabric without treatment	L	a	b	c	h°	
	32.34	−2.03	6.25	6.57	108.03	
	ΔL	Δa	Δb	Δc	Δh	ΔE
Hydroperm LPU liq c/Siligen SIH liq	0.36 L	0.00	0.43 B	−0.41 D	0.14 G	0.48
Wool Hydrolyzate KAl(SO₄)₂	0.48 L	−0.20 G	−0.24 B	−0.16 D	0.27 G	0.47

Table 10.
Color measurement of fabrics (recipe 1).

Fabric without treatment	L	a	b	c	h°	
	53.98	56.69	43.38	71.38	37.42	
	ΔL	Δa	Δb	Δc	Δh	ΔE
Wool hydrolyzate KAl(SO₄)₂	−1.21 D	−0.56 G	−0.30 B	−0.63 D	0.10 Y	0.58
Hydroperm LPU liq c/Siligen SIH liq	−1.25 D	−0.65 G	−0.12 B	−0.59 D	0.29 Y	0.62

Table 11.
Color measurement of fabrics (recipe 2).

4. Conclusion

In this study, hydrophilicity enhancer and hydrophilic silicone combinations were applied to polyamide6.6/polyurethane fabrics under two different weights employing the most used recipes of leading companies in the textile industry by using impregnation and exhaustion methods. As an alternative to the recipes, samples treated with wool hydrolysate were subjected to the same tests. The values obtained by using the exhaustion method gave better results than conventional silicone/hydrophilicity enhancer. In accordance with the firms' recommendation, 5 consecutive washes were performed, and it was observed that the values obtained with wool hydrolysate were higher.

Findings of the experiment suggests wool hydrolysate can be used instead of thermo reactive polyurethane and modified polysiloxane. These chemicals are approximately 4.5 Euro/kg and 2.5 Euro/kg, respectively. 15 L hydrolysate was obtained from 10 g of waste wool in the production of wool hydrolysate. As it is seen, it is very economical and if concentrated product is obtained instead of solution in the future, the transportation of the product will also be economical. Thus, an example of sustainable, economical and environmental work was exhibited for polyamide6.6/polyurethane blends which are used in sportswear industry regarded as an area open to development, high potential and high added value in the textile industry.

Acknowledgements

The authors thank Suleyman Demirel University Isparta, Turkey for contributions to the project and financial support: A project of Suleyman Demirel University (FYL 2019-7326). Ethics committee approval of the study was granted by the Clinical Research Ethics Committee of Suleyman Demirel University Faculty of Medicine on 30.12.2020.

Author details

Meliha Oktav Bulut* and Ayşen Cire
Department of Textile Engineering, Engineering Faculty, Suleyman Demirel University, Isparta, Turkey

*Address all correspondence to: oktavbulut@sdu.edu.tr

IntechOpen

References

[1] L. O'Connell. Size of the U.S. athletic apparel market from 2015 to 2025 Statista. [internet]. 2020. Available from: Https:/www.statista.com, Clothing and Apparel [Accessed 23 March 2020]

[2] S. Kumar and R. Deshmukh. Sports apparel market by end user (children, men, and, women) and distribution channel (E-commerce, supermarket/hypermarket, brand outlets, and discount stores): Global opportunity analysis and industry forecast, 2019-2016. [Internet]. 2020. Available from: https:/www.alliedmarketresearch.com [Accessed 23 March 2020]

[3] Martínez-Estrada, M., Moradi, B., Fernandez-Garcia, R., Gil, I. Impact of conductive yarns on an embroidery textile. Sensors. 2019;19(5):1004. DOI: 10.3390/s19051004

[4] Rauf, S., Vijjapu, M. T., Andrés, M. A., Gascón, I., Roubeau, O., Eddaoudi, M., Salama, K. N. A Highly selective metal-organic framework textile humidity sensor. ACS Appl Mater Interfaces. 2020;12(26):29999-30006. DOI:10.1021/acsami.0c07532.

[5] Manshahia, M., Das, A. High active sportswear – A critical review. Indian J. Fibre Text. Res. 2014;39:441-449.

[6] Havenith, G. 2002. Interaction of clothing and thermoregulation. Exog Dermatol. 1:221-230. DOI:10.1159/000068802

[7] Chen, Q., Miao, X., Mao, H., Ma, P., Jiang, G. The effect of knitting parameter and finishing on elastic property of pet/pbt warp knitted fabric. Autex Res. J. 2017;17(4):350-360. DOI: 10.1515/aut-2017-0014

[8] Chen, Q., Miao, X., Mao, H., Ma, P., Jiang, G. The comfort properties of two differential-shrinkage polyester warp knitted fabrics. Autex Res. J. 2016;16(2): 90-99. DOI:10.1515/aut-2015-0034

[9] Karaca, E., Kahraman, N., Ömeroğlu, S., Becerir, B. Effects of fiber cross sectional shape and weave pattern on thermal comfort properties of polyester woven fabrics. Fibres Text. East. Eur. 2012;3(92):67-72.

[10] Oğlakcioglu, N., Cay, A., Marmaralı, A. Mert, E. Characteristics of knitted structures produced by engineered polyester yarns and their blends in terms of thermal comfort. J. Eng Fibers Fabrics. 2015;10:32-41.

[11] Tomovska, E., Hes, L. Thermophysiological comfort properties of polyamide pantyhose fibres text. East. Eur. 2019;5(137):53-58. DOI: 10.5604/01.3001.0013.2902

[12] Mansor, A., Ghani, S. A., Yahya, M. F. Knitted fabric parameters in relation to comfort properties. Am. J. Mater. Sci. 2016;6(6):147-151. DOI:10.5923/j.materials.20160606

[13] Cublic, I. S., Skenderi, Z., Havenith, G. Impact of raw material, yarn and fabric parameters, and finishing on water vapor resistance Text. Res. J. 2013;83(12):1215–1228. DOI:10.1177/0040517512471745

[14] Guo, J. The Effects of Household Fabric Softeners on the Thermal Comfort and Flammability of Cotton and Polyester Fabrics [Thesis]. Virginia: Polytechnic Institute and State University, 2003.

[15] Abreu, M. J., Vidrago, C., Soares, G. M. Optimization of the thermal comfort properties of bed linen using different softening formulations. J. Text. Apparel (Tekstil ve Konfeksiyon) 2014;24:219–223.

[16] Bulut, M. O., Akbulut, Y., Halaç, E., Fidan, U., Demirezen, T. Effect of some

silicone softener on hyrophility and handle properties of woven fabrics based on PET and its blends. J. Erciyes Univ. Sci Techol. 2014;30(2):126-132.

[17] Soysal, M. I., Ünal, E. Ö. Sheep breeds genetic diversity of farm animal genetic resources of Turkiye. In proceeding of international congress on wool and luxury Fibers conference (Iconwolf). 19 April 2019, Tekirdağ.

[18] Dellal, G., Soylemezoglu, F., Erdogan, Z., Pehlivan, E., Koksal, O., Tuncer, S. S. Present situation and future of animal fiber production in Turkey: A review. J. Life Sci. 2014;8(2):192-200.

[19] Eser, B., Çelik, P., Çay, A., Akgümüş, D. Sustainability and recycling opportunities in the textile and apparel sector. Tekst. Muhendis. 2016;23(101):43-60. DOI:10.7216/1300759920162310105

[20] Khardenavis, A. A., Kapley, A., Purohit, H. J. Processing of poultry feathers by alkaline keratin hydrolyzing enzyme from *Serratia* sp. HPC 1383. Waste Manage. 2009;29(4):1409–1415. DOI:10.1016/j.wasman.2008.10.009

[21] Sinan, H. Investigation of keratin production methods from waste wools [thesis]. Isparta: Suleyman Demirel university graduate School of Applied and Natural Sciences, 2019.

[22] Sharma, S., Gupta, A. Sustainable management of keratin waste biomass: Applications and future perspectives. Braz. Arch. Biol. Technol. 2016;59:1-14.

[23] Eslahi, N., Dadashian, F., Nejad, N. H. (2013). An investigation on keratin extraction from wool and feather waste by enyzmatic hydrolysis. Prep. Biochem. Biotechnol. 2013;43(7):624-648. DOI:10.1080/10826068.2013.763826

[24] Cire, A. The finishing process of polyamide fabrics with wool

hydrolysate [thesis]. Isparta: Suleyman Demirel university graduate School of Applied and Natural Sciences, 2020.

[25] CHT, Arristan HPC T, Technical data sheet, 3 p.

[26] CHT Bezama, Tubingal SHE, Technical data sheet, 3 p.

[27] Archroma, Hydroperm LPU liq C, Technical data sheet, 2013; 4 p.

[28] Archroma, Siligen SIH liq, Technical data sheet, 2013; 5 p.

[29] Sülar, V., Okur, A. Subjective evaluation methods used in the determination of tactile properties of fabrics. Tekst. Mühendis. 2015;59-60:14-21

[30] Singh, R., Gupta, D. Moisture management in synthetic textiles. Asian Tech. Text. 2010;4:47–51.

[31] Baltusnikaite, J., Abraitiene, A., Stygiene, L., Krauledas, S., Rubeziene, V., Varnaite-Zuravliova, S. Investigation of moisture transport properties of knitted materials intended for warm underwear. Fibres Text. East. Eur. 2014;4(106):93-100.

[32] Senthilkumar, M., Sampath, M. B., Ramachandran, T. Moisture management in an active sportswear: Techniques and evaluation—A review article. J. Inst. Eng. India Ser. E 2013;93 (2):61–68.

[33] Sampath, M. B., Aruputhara, A., Senthilkumar, M., Nalankilli, G. Analysis of thermal comfort characteristics of moisture management finished knitted fabrics made from different yarns J. Ind. Text. 2012;42(1):19–33. DOI:10.1177/1528083711423952

[34] Yüksel, M. F. Investigation of a method which enable polyamide fabrics to be dyed at lower temperatures and to be printed with shorter fixation times.

Tekirdağ: Namık Kemal university graduate School of Natural and Applied Sciences Department of Textile Engineering, 2015.

[35] Rıstıć, N. N., Dodıć, I. R., Rıstıć, I. P. The influence of surfactant structure on the dyeing of polyamide knitting with acid dyes. Chemical Industry and Chemical Engineering Quarterly 2018; 24(2):117–125. DOI:10.2298/CICEQ161102025R

[36] Mullin, J. M., Sipek. M. Solubility and density isotherms for potassium aluminum sulfate-water-alcohol systems. J. Chem. Eng. Data. 1981;26:164-165. DOI:10.1021/je00024a021

[37] Udoetok, I. A., Akpanudo, N. W., Uwanta, E. J., Ubuo, E. E., Ukpong, E. J. Effect of potassium aluminium sulphate, $Kal(SO_4)_2$ on the total petroleum hydrocarbon of diesel oil. Int. J. Mat. Chem. 2012;2(3):101-104. DOI: 10.5923/j.ijmc.20120203.03

[38] Shavandi, A., Bekhit, E. A., Carne, A, Bekhit, A. Evaluation of keratin extraction from wool by chemical methods for bio-polymer application. J Bioact Compat Pol 2017;32(2):163-177.

[39] Cardamone, J. M., Nunez, A., Garcia, R. A., Aldema-Ramos, M. Characterizing wool keratin. Res. Letter. Mat. Sci. 2009:1-5. DOI:10.1155/2009/147175

[40] Balpetek, F. G., Gülümser, T. Shading of polyamide 6.6 after multi washing process and investigation of chemical changes on fabric surface. J. Text. Apparel (Tekstil ve Konfeksiyon). 2018;28(2):125-134.

[41] Mishra, A. K., Chattopadhyay, D. K., Sreedhar, B., Raju, K. V. S. N. FT-IR and XPS studies of polyurethane-urea-imide coatings. Prog. Org. Coat. 2006;55 (3):231-243. DOI:10.1016/j.porgcoat.2005.11.007

[42] Bulut, M. O., Akçalı, K. Investigation of oil removel process of the lycra blended cotton knitting fabric. J. Erciyes Univ. Sci Techol. 2012;28(3): 262-269.

[43] Parthiban, M., Ramesh-Kumar, M. Effect of fabric softener on thermal comfort of cotton and polyester fabrics. Indian J. Fibres Text. Res. 2007;32(4): 446–452.

[44] Kaplan, S., Okur, A. Thermal comfort performance of sports garments with objective and subjective measurements. Indian J. Fibre Text. Res. 2012;37(1):46-54.

[45] Kaplan, S., Okur, A. Subjective evaluation methods and physiological measurements used to determine clothing thermal comfort. 2009. 14th National Biomedical Engineering Meeting. DOI:10.1109/BIYOMUT.2009.5130345

[46] Farzana, N., Smriti, S. M. Reflectance value and yellowing propensity on thermal and storage condition of cotton fabric treated with different softeners. Int. J. Current Eng. Technol. 2015;5(1):507-511.

[47] Bulut, M. O., Akbulut, Y. The yellowing problems of top white knitting fabrics. J. Erciyes Univ. Sci Techol. 2012;28(3):233-239.

Chapter 7

Liquid Film Evaporation: Review and Modeling

Jamel Orfi and Amine BelHadj Mohamed

Abstract

Liquid film evaporation is encountered in various applications including in air humidifiers, in multiple effect distillers in thermal desalination, and in absorption cooling evaporators. It is associated with a falling pure, binary or multicomponent liquid film with associated complex and coupled heat and mass transfer processes. This chapter presents important fundamental aspects inherent to falling film evaporation in several geometrical configurations such as on horizontal tubes and inside inclined or vertical tubes or channels. The first part of the chapter concerns a review of recent works on this topic with emphasis on modeling and simulation features related to falling liquid films with heat and mass transfers. This document aims also to establish a frame for the modeling of the fluid flow with heat and mass transfer in the presence of evaporation. The main governing equations and the appropriate boundary and interfacial conditions corresponding to the fluid flow and associated heat and mass transfer and phase change are systematically presented and discussed for the case of falling film in a vertical channel with the presence of flowing gas mixture. Various simplifications of the governing equations and boundary and interfacial conditions have been proposed and justified. In particular, the formulation with extremely thin liquid film approximation is discussed.

Keywords: falling film, evaporation, evaporators, horizontal tubes, extremely thin films, modeling, thermal desalination, absorption

1. Introduction

Evaporation is a phase change process widely encountered in natural and industrial applications. Evaporation of a thin layer of alcohol or water in ambient air and evaporation of seawater film on a bundle of horizontal tubes of an evaporator are examples of such a complex phenomenon. Evaporation of liquid films occurs generally to ensure a cooling of the liquid itself, to cool the surface on which the liquid flows or to increase the concentration of some components in the liquid. The evaluation of the heat and mass transfer coefficients and associated evaporation rates in various configurations is an important task in the appropriate design and fabrication of multiple evaporators and heat exchangers needed in different applications including those related to microsystems. This explains why this topic has attracted an increasing and significant interest from the scientific and industrial communities. This chapter includes first, a review on the main recent works on the falling film evaporation and in a second phase, important fundamental aspects on modeling of the associated heat and mass transfer and fluid flow.

2. Literature review

In this section, updated literature survey gathering important studies on evaporation of single-component and multicomponent liquid films with associated transport phenomena and related applications will be presented and discussed. A focus will be on the modeling and simulations aspects of falling film evaporation systems.

2.1 Examples of applications of falling film evaporation

Falling film has been used in various applications. Two examples are given here. The first one concerns water desalination using falling liquid films. The second one is related to absorption refrigeration.

Multiple effect distillation (MED) is widely employed in thermal desalination industry as a mature and reliable technology. It is considered as best suited, compared to membrane-based desalination for treating feeds with high temperature and salinity [1].

Falling film evaporators are the core of the MED units. Feed preheated seawater is sprayed on the horizontal tubes as a falling film and is evaporated due to the latent heat of condensation of the steam circulating inside the tubes. The steam itself is condensing as a result of heat exchange with the evaporating feed water.

Extensive works have been published on modeling the fluid flow with heat and mass transfer in the evaporators of MED plants [2–5]. It is of interest to mention a recent work conducted by Jin et al. [5] on scale formation and crystallization modeling on horizontal tube falling evaporators used in MED. Jin et al. [5] reported the impact of various conditions of steam flow, seawater flow rate, and inlet temperature, and tube wall material and thickness on the main process performance parameters including the evaporate rate, scale growth, and overall heat transfer coefficient. The authors observed in particular that the scale layer thickness increases sharply as the feed water flow rate decreases or the tube steam temperature increases.

Another important application of falling film evaporators concerns cooling by absorption. In such systems, solution, such as LiBr-H2O, is sprayed over a bundle of horizontal tubes and a thin liquid film of solution is then formed around each tube. The percentage of the tubes surface covered by the liquid film known as "Wetting Ratio (WR)" is to be maximized for an efficient evaporation and cooling process. WR depends on various parameters including the mass flow rate per unit tube length, the solution surface tension, and the external tube surface roughness. Bu et al. [6] investigated experimentally and numerically the heat and mass transfer effectiveness of ammonia water in a falling film evaporation in vertical tube evaporators. The numerical model is based on the boundary layer equations of mass, quality, momentum, and energy for the binary ammonia-water system and solved by coordinate transformation. The experimental and numerical data are fairly compared for a various range of control parameters. The results show, in particular, that the inlet solution concentration has a strong influence on the heat transfer mechanism and the ammonia evaporation rate [6]. Papaefthimiou et al. [7] developed a two-dimensional model to investigate the heat and mass transfer inherent to water vapor absorption into an aqueous solution of LiBr. The numerical solution is obtained by solving the two-dimensional energy and species conservation equations using analytical expressions of the velocity components in x and y directions. Results on the impacts of various parameters including the liquid film Reynolds number and the number of tubes on the total absorption rate, solution temperature, and mass flux are presented and discussed.

2.2 Overview of falling film and associated heat and mass transfer studies

There exist exhaustive studies on the falling film liquid evaporation. The particular case of falling liquid film on horizontal tubes has been extensively investigated theoretically and experimentally [8]. This case has several advantages that include its high heat transfer rate with low film flow rates, and it involves small temperature difference and has a relatively simple structure [9]. The heat transfer coefficients in falling film evaporators are very high and can vary between 700 and 4000 W/m^{2}K depending on the evaporating solution properties [10]. Other inherent advantages of falling film evaporation include short contact time between the fluid and the heated wall, minimal pressure drop, and minimal static head [11].

Abraham and Mani [12] proposed the thermal spray coatings to enhance the convective evaporation on horizontal tube falling film evaporators. They conducted a computational flow dynamic (CFD) analysis to predict the seawater evaporation rate and heat transfer coefficient on thermal spray-coated tubes with varying roughness under vacuum conditions. The study shows that the heat transfer coefficient increases by up to 15% due to increased roughness. However, and despite other numerous attempts to enhance the overall heat transfer process, there exist several limiting operating problems such as nonuniformity of the liquid distribution over the tubes surfaces, the presence of the non-condensable gases, and high potential of fouling and scaling mainly when dealing with salty waters.

Shear stress, gravity, and surface tension are important phenomena affecting the behavior of the falling film and the effectiveness of the evaporation process. **Figure 1** illustrates the various heat transfer processes related to liquid film falling on a horizontal cylinder.

Faghri and Zhang [13] discussed important fundamental and applied features of falling film evaporators. The basic equations giving the heat and mass transfer coefficients and the evaporation rates for various cases and configurations have been compiled and discussed. Evaporation from liquid films circulating inside channels/microchannels or horizontal/inclined walls has been described, and the related phenomena have been explained. Ribatski and Jacobi [8] developed a

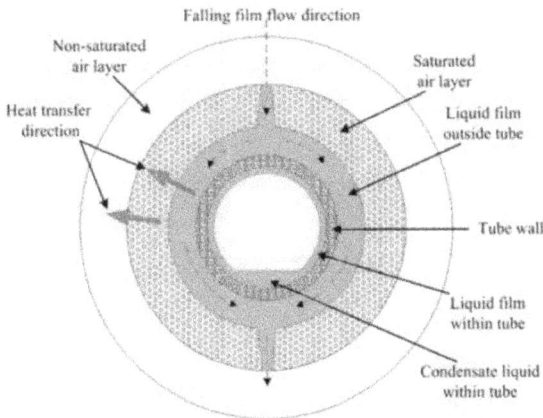

Figure 1.
Representation of heat transfer and fluid flow processes associated with falling film over a horizontal tube [9].

comprehensive and critical review on falling film evaporation on horizontal tubes. The review covers studies on heat and mass transfer performance on single tubes, finned and enhanced surfaces, and tube bundles. The authors stressed on the need to develop advanced mathematical models and accurate heat and mass transfer correlations required for the design and construction of evaporators in various applications.

The liquid film thickness and behavior are strongly linked to the heat and mass transfer coefficients and evaporation rates. It is important in the design of falling film evaporators to ensure that the film thickness is small enough to reduce the thermal resistance of the liquid layer but not too small to avoid any dry zones that may appear on the wall surface due to the rupture of the liquid which can result in various problems including fouling, corrosion, and potentially damage of the tube.

The behavior of liquid films on horizontal tubes has been investigated theoretically and experimentally in a good number of studies. It is well established that there exist three different patterns characterizing a liquid film falling over a series of horizontal tubes depending on various parameters including the liquid flow rate, the fluid properties, and the tube diameter and spacing. These flow modes are the droplet mode (the liquid leaves the tube in an intermittent way), the jet mode (the liquid leaves the tube as a continuous column), and the sheet mode (a continuous sheet is formed between the tubes) [8]. **Figure 2** describes schematically these modes.

Nusselt, as reported in [9], proposed an analytical investigation for laminar flow on horizontal tubes and one vertical or inclined wall. An expression of the film thickness by neglecting the momentum effects of the falling film was given. Similar correlation was developed by Rogers et al. [14, 15]. Later, advanced experimental methods have been used to measure the falling film thickness and characterize its patterns [10–13]. The use of these methods has led to develop clearer picture on the liquid flow structure and the associated heat and mass transfers. Besides, computational methods have been used to solve the conservation equations governing the flow and temperature fields of a falling film over surfaces [2, 14–16]. Qiu et al. [9] conducted a numerical analysis of the liquid film distribution of sheet flow on horizontal tubes. The study shows that the transient behavior of the falling film can have various stages including the free falling stage, the liquid impact stage, the liquid film developing stage, and the film fully developed stage. The presented results include the distribution of the liquid thickness with the tube diameter, the Reynolds number, and the inter-tube spacing.

Stephan [10] conducted a concise review of the heat transfer mechanisms in falling film evaporators. In particular, results and correlations on heat transfer coefficients for vertical tubes have been compiled and presented for various cases including when the falling liquid film flow is laminar, wavy laminar, and turbulent.

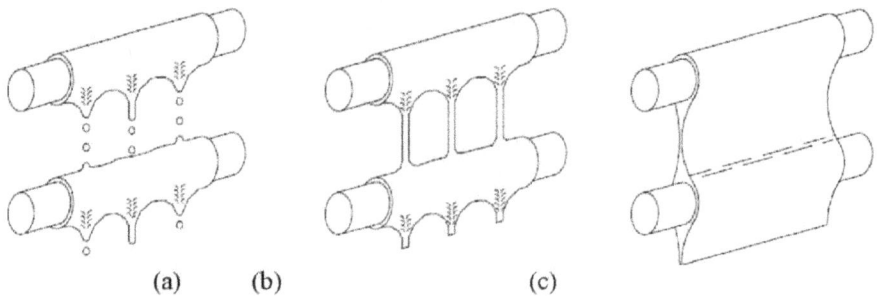

(a)   (b)   (c)

Figure 2.
The inter-tube falling film modes: (a) the droplet mode; (b) the jet mode; and (c) the sheet mode [8].

The correlations show that Nusselt number depends not only on the Reynolds and Prandtl numbers but also on Kapitza number, which measures the effect of surface tension compared to the viscous ones. Zhao et al. [17] conducted a comprehensive review on computational studies on falling liquid film flow with associated heat transfer on horizontal tubes and tube bundle. Review includes various features on falling film hydrodynamics, evaporation, and boiling outside the single tubes and the tube bundle and whole evaporator performance investigated using 2D and 3D models. Besides, previous results on falling liquid film dry-out and breakout are screened and discussed. Zhao et al. [17] concluded their review by proposing recommendations and future needs to be investigated in various fields and technologies.

There exist in general two approaches to treat numerically the heat and mass transfer associated with the evaporation of a liquid film in presence of a non-saturated gas [18–21]. The first one considers an extremely thin layer of liquid. Therefore, the governing conservation equations are simultaneously solved not only in the gas region but also in the liquid film. This requires also considering appropriate interfacial conditions between the liquid and gas phases. The second approach assumes that when the liquid film is extremely thin, the overall heat and mass transfers are not or slightly affected by the exchanges in the liquid itself. In this approach, the interfacial conditions are directly applied on the surface wall as boundary conditions. By neglecting convective terms in momentum and energy equations of the liquid, it is shown that the assumption of an extremely thin film thickness is valid only for a low mass flow rate [22]. Refs. [23–30] investigated the heat and mass transfer associated with liquid film evaporation by considering the heat transfer and fluid flow within the liquid film.

On another side, several other works have been based on the assumption of the extremely thin thickness. Cherif and Daif [21] considered the evaporation of a binary liquid film by mixed convection falling on one side of a parallel plate channel. The wetted plate is subjected to a constant and uniform heat flux, while the second one is taken as adiabatic. The authors studied the impact of using the very thin film assumption on the heat and mass transfer results. They showed in particular that the overestimation induced by considering an extremely thin film is greater for the ethylene/glycol-water mixture than for the ethanol-water mixture.

Recently, Alami et al. [31] studied the evaporative heat and mass transfer of a turbulent falling liquid film in a finite vertical tube that is partially heated. Using an implicit finite difference model, the authors solved the governing mass, species, momentum, and energy equations considering appropriate boundary and interfacial conditions. The obtained data are compared to the case of the entirely heated tube wall. Belhadj Mohamed and Tlili [32] analyzed the evaporation of a seawater film by mixed convection of humid air. In another study, Belhadj Mohamed et al. [33] considered the impact of adding metal nanoparticles to the falling liquid on the effectiveness of the evaporation process. Ma et al. [34] presented a novel model to investigate the flow and evaporation of liquid film in a rocket combustion chamber with high temperature and high shear force.

In addition to the theoretical and numerical studies on falling film evaporation, the literature includes extensive experimental research activities [35–37]. Yue et al. [35] designed and conducted a series of experiments to analyze the falling film flow behavior and evaluate the associated heat transfer outside a vertical tube. New correlations on the heat transfer coefficient and falling film dry burning have been proposed. Shahzad et al. [37] considered practical features related to the design of industrial falling film evaporators. They enumerated the main advantages of these types of evaporators and reviewed the corresponding heat transfer correlations.

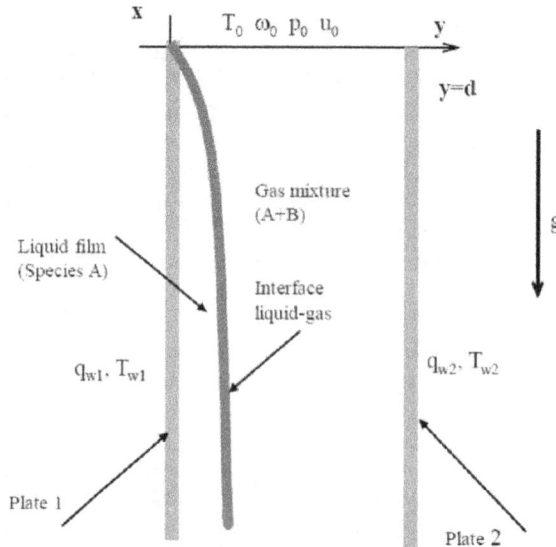

Figure 3.
Representation of the physical model.

Besides, they conducted an experimental study and proposed their own falling film heat transfer correlation.

3. Modeling of the heat and mass transfer with falling liquid film in confined channels

3.1 Introduction

We present in this section some aspects related to the theoretical formulation of the heat and mass transfer associated with liquid falling film in confined domains. We will be limited to the laminar steady state nature of flows and to the two-dimensional Cartesian configuration. We will give particular interest to the interfacial conditions equations.

3.2 Physical model description

We consider the flow of a thin liquid falling film on a plate of a vertical channel with the presence of a binary mixture gas flow. The gas and liquid flows are supposed laminar and in steady state regime. The gas mixture is composed of a non-condensable chemical species B with high concentration and a species A as vapor. This mixture can be, for example, a humid air mixture or an air-alcohol mixture. **Figure 3** shows schematically the system under study which can represent a heat and mass exchanger between a liquid film and a gas in direct contact. Various phenomena characterize this system such as thin liquid evaporation, vapor condensation, and shear stress between the gas and liquid flows. In addition, the difference in concentration that can exist between the liquid-gas interface, supposed saturated in species A as a vapor, and the neighboring gaseous mixture may result in a diffusion of the component A from the interface to the gas in case of evaporation or a reverse diffusion (from the gas toward the interface) in the condensation case.

It is worthy to mention that in absence of a forced flow of the gas mixture, the natural flow induced by the temperature and concentration gradients within the gas can be upward or downward depending on the two gases and the subjected heat and mass conditions.

3.3 Governing equations

The equations governing the flows and transfers in the two phases are those of mass, momentum, and energy equations. For laminar, steady state with no chemical reactions and neglecting the radiative heat transfer in the two fluids, the viscous dissipation and the pressure work, the conservation equations are as follows [38, 39]

3.3.1 Continuity equation

$$\frac{\partial}{\partial x}(\rho u) + \frac{\partial}{\partial y}(\rho v) = 0 \tag{1}$$

3.3.2 Momentum equations

- in x direction:

$$\rho\left(u\frac{\partial u}{\partial x} + v\frac{\partial u}{\partial y}\right) = -\frac{\partial p}{\partial x} + \frac{\partial}{\partial x}\left(2\mu\frac{\partial u}{\partial x} - \frac{2}{3}\mu\nabla\vec{V}\right) + \frac{\partial}{\partial y}\left(\mu\left(\frac{\partial u}{\partial y} + \frac{\partial v}{\partial x}\right)\right) + \rho g_x \tag{2}$$

- in y direction:

$$\rho\left(u\frac{\partial v}{\partial x} + v\frac{\partial v}{\partial y}\right) = -\frac{\partial p}{\partial y} + \frac{\partial}{\partial y}\left(2\mu\frac{\partial v}{\partial y} - \frac{2}{3}\mu\nabla\vec{V}\right) + \frac{\partial}{\partial x}\left(\mu\left(\frac{\partial u}{\partial y} + \frac{\partial v}{\partial x}\right)\right) + \rho g_y \tag{3}$$

where ρ and μ are the fluid density and dynamic viscosity, respectively; u and v are x and y components of the velocity $\vec{V}$; and p is the total pressure, while g_x and g_y are the x and y gravity acceleration components, respectively.

The total pressure can be written as the summation of the hydrostatic pressure p_0 and the dynamic pressure $(p - p_0)$. The hydrostatic pressure can be expressed as:

$$\frac{\partial p_0}{\partial x} = \rho_0 g_x = -\rho_0 g \tag{4}$$

The term $-\frac{\partial p}{\partial x} + \rho g_x$ in the Eq. (2) can be written as:

$$-\frac{\partial p}{\partial x} + \rho g_x = -\frac{\partial p}{\partial x} - \rho g = -\frac{\partial(p-p_0)}{\partial x} + (\rho_0-\rho)g \tag{5}$$

p_0 is the fluid density at the reference 0. The quantity $(\rho_0-\rho)g$ refers to natural convection generation. For small variations within the thermal and concentration fields, $(\rho_0-\rho)$ can be expressed as function of the temperature and the concentration using the Boussinesq approximation as [39]:

In the liquid

$$(\rho_0 - \rho) = \rho_0 \beta_l (T - T_0) \tag{6}$$

β_l refers to the thermal expansion coefficient in the liquid.

In the gaseous mixture

$$(\rho_0 - \rho) = \rho_0(\beta_T(T - T_0) + \beta_w(w - w_0)) \tag{7}$$

β_T and β_w are the thermal expansion and the mass expansion coefficients, respectively. w is the mass concentration of constituent A.

When the gas behaves as an ideal gas, β_T and β_w can be expressed respectively as:

$$\beta_T = \frac{1}{T_0} \tag{8}$$

and

$$\beta_\omega = \left(\omega_{A0} + \frac{M_A}{M_B - M_A}\right)^{-1} \tag{9}$$

M_A and M_B are the molar mass of constituent A (minority constituent) and constituent B (non-condensable majority constituent B).

For a binary ideal gas,

$$\rho = \frac{pM}{RT}, \rho_0 = \frac{p_0 M_0}{RT_0} \tag{10}$$

and $(\rho - \rho_0)$ becomes

$$(\rho - \rho_0) = \rho\left(1 - \frac{\rho_0}{\rho}\right) = \rho\left(1 - \frac{p_0 M_0 T}{pMT_0}\right) \tag{11}$$

M stands for mixture molar mass.

The pressure variation is considered much smaller than the molar mass or temperature [39]. Then, we can have

$$\left(1 - \frac{\rho_0}{\rho}\right) \approx \left(1 - \frac{M_0}{M}\frac{T}{T_0}\right) \tag{12}$$

The left term in this equation can be expressed as:

$$\left(1 - \frac{\rho_0}{\rho}\right) \approx \left(1 - \frac{\rho_0}{\rho}\right)_{T=const} + \left(1 - \frac{\rho_0}{\rho}\right)_{w=const} \tag{13}$$

or:

$$\left(1 - \frac{M_0}{M}\frac{T}{T_0}\right) \approx \left(1 - \frac{M_0}{M}\right) + \left(1 - \frac{T}{T_0}\right) \tag{14}$$

The mixture molar mass can be expressed in terms of the molar mass of constituents A and B and their mass concentrations ω_A and ω_B as:

$$M = \frac{M_A M_B}{\omega_A M_B + \omega_B M_A} \tag{15}$$

Then:

$$\left(1 - \frac{M_0}{M}\right) \approx \left(1 - \frac{\omega_A M_B + (1 - \omega_A)M_A}{\omega_{A0}M_B + (1 - \omega_{A0})M_A}\right) \tag{16}$$

or

$$\left(1 - \frac{M_0}{M}\right) \approx \left(\frac{\omega_{A0} - \omega_A}{\omega_{A0} + \frac{M_A}{M_B - M_A}}\right) \tag{17}$$

Finally:

$$(\rho - \rho_0) = \rho \left[\left(1 - \frac{T}{T_0}\right) + \frac{\omega_{A0} - \omega_A}{\left(\omega_{A0} + \frac{M_A}{M_B - M_A}\right)} \right]$$

$$= -\rho \left[\left(\frac{T - T_0}{T_0}\right) + (\omega_A - \omega_{A0}) \left(\omega_{A0} + \frac{M_A}{M_B - M_A}\right)^{-1} \right]$$

or

$$(\rho_0 - \rho) = \rho[\beta_T(T - T_0) + \beta_\omega(\omega_A - \omega_{A0})] \tag{18}$$

where

$$\beta_T = \frac{1}{T_0} \text{ and } \beta_\omega = \left(\omega_{A0} + \frac{M_A}{M_B - M_A}\right)^{-1} \tag{19}$$

When one neglects ω_{A0} as compared to $\frac{M_A}{M_B - M_A}$, β_ω can be written as:

$$\beta_\omega \approx \frac{M_B - M_A}{M_A} \tag{20}$$

3.3.3 Energy conservation equation

$$\rho C_p \left(u \frac{\partial T}{\partial x} + v \frac{\partial T}{\partial y}\right) = -\frac{\partial q_x}{\partial x} - \frac{\partial q_y}{\partial y} \tag{21}$$

q_x and q_y are the x and y components of the heat flux $\vec{q}$.

3.3.4 Species conservation equation

$$\rho \left(u \frac{\partial \omega_A}{\partial x} + v \frac{\partial \omega_A}{\partial y}\right) = -\frac{\partial J_{Ax}}{\partial x} - \frac{\partial J_{Ay}}{\partial y} \tag{22}$$

J_{Ax} and J_{Ay} are the x and y mass flux components of species A with respect to average mixture velocity.

3.4 Soret and Dufour interdiffusion effects

The mass and heat fluxes $\vec{J_A}$ and $\vec{q}$, respectively, depend on the concentration and temperature gradients. They are expressed as [39–41]:

$$\vec{q} = -k\overrightarrow{grad}T + \left(\alpha_d RT \frac{M^2}{M_A M_B} + (h_A - h_B)\right)\vec{J_A} \tag{23}$$

$$\vec{J_A} = -\rho D_{AB}\left(\overrightarrow{grad}\omega_A + \alpha_d \omega_A (1 - \omega_A)\overrightarrow{grad}(\ln T)\right) \tag{24}$$

α_d is a thermal diffusion factor, R is the universal gas constant, h is the specific enthalpy, k is the thermal conductivity, and D_{AB} is the coefficient of diffusion of species A in the mixture (A + B). The second and third terms of the equation giving $\vec{q}$ Eq. (23) refer to the contribution associated with the concentration gradient (Dufour effect) and with the interdiffusion of species. The second term of the equation giving the mass flux $\vec{J_A}$ Eq. (24) refers to the temperature gradient (Soret effect).

The Dufour and Soret are neglected in the majority of studies on coupled heat and mass transfers [34]. Gebhart et al. [39] reported that these effects can be neglected when the molar masses of the constituents are close and the variations in the concentration of the diffusing species are not significant. The interdiffusion of species becomes important when the difference between the specific heat coefficients of species A and B is high [41].

After substitution and adjustment, the energy and species conservation equations become

$$\rho C_p \left(u \frac{\partial T}{\partial x} + v \frac{\partial T}{\partial y} \right) = \frac{\partial}{\partial x} \left(k \frac{\partial T}{\partial x} \right) + \frac{\partial}{\partial y} \left(k \frac{\partial T}{\partial y} \right) - \frac{\partial}{\partial x} \left[\left(\frac{R}{M_A M_B} (\alpha_d M^2 T) \right) \right.$$

$$\left. + (h_A - h_B) J_{Ax} \right] - \frac{\partial}{\partial y} \left[\left(\frac{R}{M_A M_B} (\alpha_d M^2 T) \right) + (h_A - h_B) J_{Ay} \right]$$

$$(25)$$

$$\rho \left(u \frac{\partial \omega_A}{\partial x} + v \frac{\partial \omega_A}{\partial y} \right) = \frac{\partial}{\partial x} \left(\rho D_{AB} \frac{\partial \omega_A}{\partial x} \right) + \frac{\partial}{\partial y} \left(\rho D_{AB} \frac{\partial \omega_A}{\partial y} \right)$$

$$+ \frac{\partial}{\partial x} \left[\rho D_{AB} \alpha_d \omega_A (1 - \omega_A) \frac{1}{T} \frac{\partial T}{\partial x} \right] + \frac{\partial}{\partial y} \left[\rho D_{AB} \alpha_d \omega_A (1 - \omega_A) \frac{1}{T} \frac{\partial T}{\partial y} \right]$$

$$(26)$$

3.5 Boundary conditions

Different types of thermal, mass, and hydrodynamic boundary conditions relating to the physical system shown schematically in **Figure 3** can be considered. Thus and by way of illustration, we consider the situation where the two plates of the channel in **Figure 3** are subjected to constant heat fluxes q_{w1} and q_{w2}. Plate 2 is impermeable and dry. This translates into:

- On plate 1 (y = 0), one can write

$$q_{w1} = q_y \Big)_{y=0} = -k_l \frac{\partial T_l}{\partial y} \Big)_{y=0} \qquad (27)$$

When the liquid film thickness is negligible, one can have

$$q_{w1} = q_y \Big)_{y=0} = -k \frac{\partial T}{\partial y} \Big)_{y=0} + \left[\alpha_d \frac{RTM^2}{M_A M_B} + (h_A - h_B) \right] J_{Ay} \Big)_{y=0} + q_l \qquad (28)$$

q_l refers to the latent heat transfer.

For the mass transfer, the saturation is translated by:

$$\omega \Big)_{y=0} = \omega_{sat}(T(x, 0)) \qquad (29)$$

ω_{sat} stands for the saturated vapor concentration.

- Plate 2 is impermeable. The mass flux is expressed as:

$$J_{Ay}\Big)_{y=d} = -\rho D_{AB}\left[\left(\frac{\partial \omega}{\partial y} + \alpha_d \frac{\omega_A(1-\omega_A)}{T}\frac{\partial T}{\partial y}\right)\Big|_{y=d}\right] = 0 \qquad (30)$$

Therefore, the diffusion mass transfer is balanced by that associated with the Soret effect.

The thermal boundary condition is reduced in this case to:

$$q_{w2} = q_y\Big)_{y=d} = -k\frac{\partial T}{\partial y}\Big)_{y=d} \qquad (31)$$

- On another side, the nonslip and impermeability conditions are expressed as follows:

- On plate 1 (y = 0),

$$u_l(x,0) = v_l(x,0) = 0 \qquad (32)$$

- On plate 2 (y = d),

$$u(x,d) = v(x,d) = 0 \qquad (33)$$

It is worthy to mention that when the liquid film thickness is extremely small, the normal velocity on the plate is not zero. It can be obtained by applying a mass balance on the pas-wall interface. Let v_A, v_B and v be the local velocities of species A, B, and mixture (A + B), respectively, with respect to fixed reference. Also, $\dot{m}_A$ and $\dot{m}_B$ are the mass fluxes of A and B with respect to fixed reference.

$$\dot{m} = \dot{m}_A + \dot{m}_B \text{ or } \rho v = \rho_A v_A + \rho_B v_B \qquad (34)$$

The diffusion mass flux of A, J_A is given by:

$$J_A = \rho_A(v_A - v) \qquad (35)$$

then

$$\dot{m}_A = J_A + \rho_A v = J_A + \frac{\rho_A}{\rho}(\rho_A v_A + \rho_B v_B) \qquad (36)$$

The interface is supposed impermeable to species B. The mass flux of B $\dot{m}_B$ is then zero on the interface. We have

$$\dot{m}_A(1-\omega_A) = J_A \Rightarrow \dot{m}_A = \rho v = \frac{J_A}{(1-\omega_A)} \qquad (37)$$

Therefore

$$v = \frac{J_A}{\rho(1-\omega_A)}\Big)_{y=0} \qquad (38)$$

This interfacial velocity is not known a priori because it depends on the concentration and temperature gradients at this location.

3.6 Liquid-gas interfacial conditions

3.6.1 General condition at the interface of two fluids

The fluid flow governing equations can be expressed in the following general form of a transport equation:

$$\frac{\partial \phi}{\partial t} + div \vec{f} = S \qquad (39)$$

Also on an integral form:

$$\frac{d}{dt} \int_\tau \phi d\tau + \int_A \vec{f}.\vec{n}dA = \int_\tau Sd\tau \qquad (40)$$

φ denotes the volume density of any physical quantity. $\vec{f}$ is the flux of this quantity and S is the source term. τ and A refer, respectively, to the control volume and to the control surface considered. For the continuity equation, for example, we have.

$$\phi = \rho, \vec{f} = \rho\vec{v}, S = 0$$

Hsieh and Ho [42] considered a fixed control volume between two fluids (fluid 1 and fluid 2) as shown in **Figure 4**. This is a base cylinder B and height L. The height above the mobile interface is denoted by L1.

Let the mathematical function $F\left(\vec{x}, t\right)$ define the interface. $F\left(\vec{x}, t\right)$ is positive in the region of fluid 1, negative for region of fluid 2, and zero on the interface.

Applying the general Eq. (40) on the control volume of **Figure 4**:

$$\frac{d}{dt}[\phi_1 BL_1 + \phi_2 B(L - L_1)] + \left[\vec{f}_1.\frac{\overrightarrow{gradF}}{|\overrightarrow{gradF}|} - \vec{f}_2.\frac{\overrightarrow{gradF}}{|\overrightarrow{gradF}|}\right]B = SLB + o(L) \qquad (41)$$

o(L) refers to the flux through the lateral sides of the cylinder.

The source term can be composed of a volumetric source S_v and a surface source S_s [42, 43]:

$$SL = S_v L + S_s \qquad (42)$$

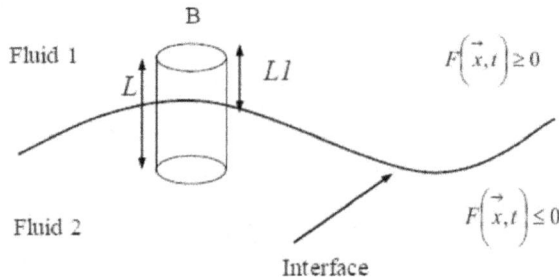

Figure 4.
Control volume at the interface between two fluids.

When L tends toward 0, Eq. (41) becomes

$$(\phi_1 - \phi_2)\frac{dL_1}{dt} + \left(\vec{f}_1 - \vec{f}_2\right)\frac{\overrightarrow{gradF}}{\left|\overrightarrow{gradF}\right|} = S_s \tag{43}$$

Consider a point M on the interface. $\vec{n}_s$ is the normal vector to the surface on M. $d\left(\overrightarrow{OM}\right)$ is the change of $\overrightarrow{OM} = \vec{x}$. The variation on $\vec{n}_s$ is $\vec{n}_s.d\left(\overrightarrow{OM}\right)$. On the interface, $F\left(\vec{x}, t\right) = 0$. Then, one can have

$$\frac{dL_1}{dt} = -\vec{n}_s.\frac{d\vec{x}}{dt} = -\frac{\overrightarrow{gradF}}{\left|\overrightarrow{gradF}\right|}\cdot\frac{d\vec{x}}{dt} \tag{44}$$

and

$$dF = \frac{\partial F}{\partial t}dt + \overrightarrow{gradF}.d\vec{x} = 0 \tag{45}$$

Combining Eqs. (44) and (45), one gets

$$\frac{dL_1}{dt} = \frac{\frac{\partial F}{\partial t}}{\left|\overrightarrow{gradF}\right|} \tag{46}$$

After substitution, a general condition expressing the conservation equation on the interface between the two fluids (1) and (2) can be obtained as:

$$\phi_1\frac{\partial F}{\partial t} + \vec{f}_1.\overrightarrow{gradF} = \phi_2\frac{\partial F}{\partial t} + \vec{f}_2.\overrightarrow{gradF} + S_s\left|\overrightarrow{gradF}\right| \tag{47}$$

In the following, we give some examples on how to apply this general equation to develop the mass, momentum, and energy conservation equations on the interface of two fluids. We consider the case where the fluid 1 is a binary gas mixture and the fluid 2 is a homogeneous liquid.

The function F can be chosen such as:

$$F = y - \delta \tag{48}$$

$\delta = \delta(x, t)$ is the liquid film thickness; x and y are the axial and transversal coordinates, respectively; y is measured from the wall on which the liquid flows.

Table 1 compiles the expressions of $\phi, \vec{f}$, and S quantities associated with conservation equations of mass, momentum, energy, and species.

- The indices i and j refer to x and y coordinates, respectively.

- e stands for the internal energy and V is the magnitude of the velocity $\vec{V}$.

- σ_{ij} refers to the normal and tangential constraints. They are expressed for a Newtonian fluid [40] as:

$$\begin{cases} \sigma_{ii} = -p - \frac{2}{3}\mu.\text{div}\left(\vec{V}\right) + 2\mu\frac{\partial u_i}{\partial x_i} \\ \sigma_{ij} = \mu\left(\frac{\partial u_i}{\partial x_j} + \frac{\partial u_j}{\partial x_i}\right) \\ \text{when } i \neq j \end{cases} \tag{49}$$

Conservation equation	φ	$\vec{f}$	S_s
Mass	ρ	$\rho\vec{v}$	0
Momentum	ρv_i	$f_j = \rho v_i v_j - \sigma_{ij}$	B_i
Energy	$\rho\left(e + \frac{V^2}{2}\right)$	$f_i = \rho(e + V^2/2)v_i$ $-\sigma_{ij}v_j + q_i + g_i(\gamma)$	Q
Species	$\rho\omega_A$	$f_i = \rho\omega_A v_i + j_{Ai}$	0

Table 1.
Expressions of ϕ, $\vec{f}$, and S as function of conservation equations (compiled, adjusted and adapted from [42–44].

- B_i is the surface source term in the momentum equation. It is given by [42, 43]:

$$\vec{B} = -\vec{n}\,div\left(\gamma\vec{n}\right) + \overrightarrow{grad}\gamma \qquad (50)$$

$\vec{n}$ is the normal vector of the interface at the point M. γ is the coefficient of surface tension. Eq. (50) shows the superposition of the tangential and normal effects of the surface tension.

- q_i and J_{Ai} represent the heat and mass fluxes on i direction.

- Q is the source term in the energy equation.

- g_i (γ) refers to the energy quantity associated with surface tension work.

3.6.2 Continuity equation case

For the conservation of mass equation, the physical quantity of interest is the mass. $\phi, \vec{f}$ and S become ρ, $\rho\vec{v}$, and 0, respectively, as shown in **Table 1**. The interfacial general Eq. (47) becomes

$$\rho_g\left(\frac{\partial\delta}{\partial t} + u_g\frac{\partial\delta}{\partial x} - v_g\right) = \rho_l\left(\frac{\partial\delta}{\partial t} + u_l\frac{\partial\delta}{\partial x} - v_l\right) = -\xi \qquad (51)$$

The indices g and l refer, respectively, to the gas and liquid.
Under steady state conditions neglecting the liquid film thickness variation, one can get

$$\rho_g v_g = \rho_l v_l \qquad (52)$$

This equation states that the mass flow rate of the gas (fluid 1) leaving (arriving to) the interface is equal to the mass flow rate of the liquid (fluid 2) which arrives to (leaves) the interface.

3.6.3 Conservation of species equation case

For the case of a liquid film (fluid 2) in contact with a nonreactive gas mixture (fluid 1) composed of A (minority species) and B (noncondensable majority species), quantities ϕ and $\vec{f}$ and S are as follows:

for the liquid

$$\phi = \rho, \vec{f} = \rho\left(u\vec{i} + v\vec{j}\right) \text{ and } S = 0 \tag{53}$$

for the gas

$$\phi = \rho\omega_A, \vec{f} = \left(\rho\omega_A u + J_{Ax}\right)\vec{i} + \left(\rho\omega_A v + J_{Ay}\right)\vec{j} \text{ and } S = 0 \tag{54}$$

The general equation on the liquid-gas interface becomes under these conditions:

$$\rho_g\omega_A\left(-\frac{\partial\delta}{\partial t}\right) + \left[\left(\rho_g\omega_A u_g + j_{Ax}\right)\left(-\frac{\partial\delta}{\partial x}\right) + \left(\rho_g\omega_A v_g + J_{Ay}\right)\right]$$
$$= \rho_l\left(-\frac{\partial\delta}{\partial t}\right) + \rho_l u_l\left(-\frac{\partial\delta}{\partial x}\right) + \rho_l v_l \tag{55}$$

For steady state regime, one can have

$$\left[\left(\rho_g\omega_A u_g + j_{Ax}\right)\left(-\frac{\partial\delta}{\partial x}\right) + \left(\rho_g\omega_A v_g + J_{Ay}\right)\right] = \rho_l u_l\left(-\frac{\partial\delta}{\partial x}\right) + \rho_l v_l \tag{56}$$

When in addition the liquid film thickness varies very little, one can get

$$\rho_g\omega_A v_g + J_{Ay} = \rho_l v_l \tag{57}$$

Under these conditions, we have also based on Eq. (52) $\rho_g v_g = \rho_l v_l$,
or

$$v_g = \frac{J_{Ay}}{\rho_g(1 - w_A)} \tag{58}$$

J_{Ay} can be given by the Fick's law, the gas velocity on the interface is expressed as:

$$v_g = -\frac{D_{AB}}{(1 - w_A)}\frac{\partial w_A}{\partial y} \tag{59}$$

3.6.4 Momentum equation case

In this case, the variable of interest is the momentum equation in the i direction. The variables ϕ and $\vec{f}$ can be expressed as:
In the x direction:

$$\phi = \rho u \text{ and } \vec{f} = \left(\rho u u - \sigma_{xx}\right)\vec{i} + \left(\rho u v - \sigma_{xy}\right)\vec{j} \tag{60}$$

In the y direction:

$$\phi = \rho v \text{ and } \vec{f} = \left(\rho u v - \sigma_{xy}\right)\vec{i} + \left(\rho v v - \sigma_{yy}\right)\vec{j} \tag{61}$$

σ_{xx}, σ_{yy}, and σ_{xy} are the normal and tangential constraints for a Newtonian fluid. They are expressed in Eq. (49).

The surface source term S_s, which is related to the surface tension effects, is given by Eq. (50). The surface tension coefficient γ can vary along the interface if this interface is nonhomogeneous for example. For a homogeneous interface, this coefficient can be taken as constant. Eq. (50) becomes

$$\vec{S}_s = -\gamma \vec{n}\, div\left(\vec{n}\right) \tag{62}$$

$\vec{n}$ is the normal vector to the fluid 2 surface:
Given that

$$div\left(\vec{n}\right) = div\left(\frac{\overrightarrow{gradF}}{\left|\overrightarrow{gradF}\right|}\right) = \left(\frac{1}{R1} + \frac{1}{R2}\right) \tag{63}$$

where R1 and R2 are the radii of curvature of the interface.
Therefore, the general equation at the interface (16) becomes the momentum conservation case:
in x:

$$\rho_g u_g\left(-\frac{\partial \delta}{\partial t}\right) + \left[\left(\rho_g u_g u_g - \sigma_{xx,g}\right)\left(-\frac{\partial \delta}{\partial x}\right) + \left(\rho_g u_g v_g - \sigma_{xy,g}\right)\right]$$
$$= \rho_l u_l\left(-\frac{\partial \delta}{\partial t}\right) + \left[\left(\rho_l u_l u_l - \sigma_{xx,l}\right)\left(-\frac{\partial \delta}{\partial x}\right) + \left(\rho_l u_l v_l - \sigma_{xy,l}\right)\right] \tag{64}$$
$$-\gamma\left(\frac{1}{R1} + \frac{1}{R2}\right)\left(-\frac{\partial \delta}{\partial x}\right)$$

in y:

$$\rho_g v_g\left(-\frac{\partial \delta}{\partial t}\right) + \left[\left(\rho_g u_g v_g - \sigma_{xy,g}\right)\left(-\frac{\partial \delta}{\partial x}\right) + \left(\rho_g v_g v_g - \sigma_{yy,g}\right)\right]$$
$$= \rho_l v_l\left(-\frac{\partial \delta}{\partial t}\right) + \left[\left(\rho_l u_l v_l - \sigma_{xy,l}\right)\left(-\frac{\partial \delta}{\partial x}\right) + \left(\rho_l v_l v_l - \sigma_{yy,l}\right)\right] \tag{65}$$
$$-\gamma\left(\frac{1}{R1} + \frac{1}{R2}\right)$$

Or after substitution and arrangement:
in x:

$$u_g\zeta + \left[-P_g - \frac{2}{3}\mu_g\left(\frac{\partial u_g}{\partial x} + \frac{\partial v_g}{\partial y}\right) + 2\mu_g\frac{\partial u_g}{\partial x}\right]\left(\frac{\partial \delta}{\partial x}\right) - \mu_g\left(\frac{\partial u_g}{\partial y} + \frac{\partial v_g}{\partial x}\right) =$$
$$u_l\zeta + \left[-P_l - \frac{2}{3}\mu_l\left(\frac{\partial u_l}{\partial x} + \frac{\partial v_l}{\partial y}\right) + 2\mu_l\frac{\partial u_l}{\partial x}\right]\left(\frac{\partial \delta}{\partial x}\right) - \mu_l\left(\frac{\partial u_l}{\partial y} + \frac{\partial v_l}{\partial x}\right) \tag{66}$$
$$+\gamma\left(\frac{1}{R1} + \frac{1}{R2}\right)\left(\frac{\partial \delta}{\partial x}\right)$$

in y:

$$v_g \zeta + \left(\mu_g \left(\frac{\partial u_g}{\partial y} + \frac{\partial v_g}{\partial x} \right) \right) \left(\frac{\partial \delta}{\partial x} \right) - \left[-p_g - \frac{2}{3} \mu_g \left(\frac{\partial u_g}{\partial x} + \frac{\partial v_g}{\partial y} \right) + 2\mu_g \frac{\partial v_g}{\partial y} \right] =$$

$$v_l \zeta + \left(\mu_l \left(\frac{\partial u_l}{\partial y} + \frac{\partial v_l}{\partial x} \right) \right) \left(\frac{\partial \delta}{\partial x} \right) - \left[-p_l - \frac{2}{3} \mu_l \left(\frac{\partial u_l}{\partial x} + \frac{\partial v_l}{\partial y} \right) + 2\mu_l \frac{\partial v_l}{\partial y} \right] \qquad (67)$$

$$- \gamma \left(\frac{1}{R1} + \frac{1}{R2} \right)$$

ζ is defined in Eq. (51).

Eqs. (66) and (67) express the momentum conservation on the liquid-gas interface in presence of phase change (evaporation/condensation) and surface tension. They can lead to easier expressions depending on the approximations considered.

Thus, for steady state and neglected liquid thickness, one can get in x direction:

$$\rho_g u_g v_g - \mu_g \left(\frac{\partial u_g}{\partial y} + \frac{\partial v_g}{\partial x} \right) = \rho_l u_l v_l - \mu_l \left(\frac{\partial u_l}{\partial y} + \frac{\partial v_l}{\partial x} \right) \qquad (68)$$

Considering Eq. (52) and the nonslip condition between the two phases (liquid-gas) ($u_g = u_l$) Eq. (68) becomes

$$\mu_g \left(\frac{\partial u_g}{\partial y} + \frac{\partial v_g}{\partial x} \right) = \mu_l \left(\frac{\partial u_l}{\partial y} + \frac{\partial v_l}{\partial x} \right) \qquad (69)$$

in y direction:

$$\rho_g v_g v_g - \left[-p_g - \frac{2}{3} \mu_g \left(\frac{\partial u_g}{\partial x} + \frac{\partial v_g}{\partial y} \right) + 2\mu_g \frac{\partial v_g}{\partial y} \right] =$$

$$\rho_l v_l v_l - \left[-p_l - \frac{2}{3} \mu_l \left(\frac{\partial u_l}{\partial x} + \frac{\partial v_l}{\partial y} \right) + 2\mu_l \frac{\partial v_l}{\partial y} \right] - \gamma \left(\frac{1}{R1} + \frac{1}{R2} \right) \qquad (70)$$

We can observe that in these conditions, the surface tension effects appear just in y direction.

When the boundary layer approximations are used neglecting the surface tension effects, the following equations widely adopted in the literature are obtained:

$$\mu_g \left(\frac{\partial u_g}{\partial y} \right) = \mu_l \left(\frac{\partial u_l}{\partial y} \right) \qquad (71)$$

$$p_g = p_l \qquad (72)$$

3.6.5 Conservation of energy equation case

For the energy equation, the variables ϕ and $\vec{f}$ should correspond to the followings (see **Table 1**):

$$\phi = \rho \left(e + \frac{V^2}{2} \right) \qquad (73)$$

$$\vec{f} = \left[\rho\left(e + \frac{V^2}{2}\right)u - \sigma_{xx}u - \sigma_{xy}v + q_x + g_x(\gamma) \right]\vec{i} \tag{74}$$
$$+ \left[\rho\left(e + \frac{V^2}{2}\right)v - \sigma_{yx}u - \sigma_{yy}v + q_y + g_y(\gamma) \right]\vec{j}$$

When the source term is zero and the work associated with the surface tension forces is neglected, the use of the general Eq. (47) gives

$$\left(e_g + \frac{V_{g2}}{2}\right)\zeta + \left\{ \left[-p_g - \frac{2}{3}\mu_g\left(\frac{\partial u_g}{\partial x} + \frac{\partial v_g}{\partial y}\right) + 2\mu_g\frac{\partial u_g}{\partial x} \right]u_g \right.$$

$$\left. + \left[\left(\mu_g\left(\frac{\partial u_g}{\partial y} + \frac{\partial v_g}{\partial x}\right)\right)v_g - q_{xg} \right\} \left(\frac{\partial \delta}{\partial x}\right) - \left[\left(\mu_g\left(\frac{\partial u_g}{\partial y} + \frac{\partial v_g}{\partial x}\right)\right)u_g \right.\right.$$

$$\left. + \left[-p_g - \frac{2}{3}\mu_g\left(\frac{\partial u_g}{\partial x} + \frac{\partial v_g}{\partial y}\right) + 2\mu_g\frac{\partial v_g}{\partial y}\right]v_g - q_{yg} \right] =$$

$$\left(e_l + \frac{V_{l2}}{2}\right)\zeta + \left\{ \left[-p_l - \frac{2}{3}\mu_l\left(\frac{\partial u_l}{\partial x} + \frac{\partial v_l}{\partial y}\right) + 2\mu_l\frac{\partial u_l}{\partial x}\right]u_l + \left[\left(\mu_l\left(\frac{\partial u_l}{\partial y} + \frac{\partial v_l}{\partial x}\right)\right)\right]v_l \right.$$

$$\left. - q_{x,l} \right\} \left(\frac{\partial \delta}{\partial x}\right) - \left[\left(\mu_l\left(\frac{\partial u_l}{\partial y} + \frac{\partial v_l}{\partial x}\right)\right)u_l + \left[-p_l - \frac{2}{3}\mu_l\left(\frac{\partial u_l}{\partial x} + \frac{\partial v_l}{\partial y}\right) + 2\mu_l\frac{\partial v_l}{\partial y}\right]v_l \right.$$

$$\left. - q_{y,l} \right] \tag{75}$$

If in addition, the flow is steady state and the work of the friction forces and kinetic energy terms are ignored, and Eq. (75) becomes

$$\left(e_g + \frac{p_g}{\rho_g}\right)\left[-\rho_g\left(u_g\frac{\partial \delta}{\partial x} - v_g\right) \right] + q_{y,g} - q_{xg}\frac{\partial \delta}{\partial x}$$

$$= \left(e_l + \frac{p_l}{\rho_l}\right)\left[-\rho_l\left(u_l\frac{\partial \delta}{\partial x} - v_l\right) \right] + q_{y,l} - q_{x,l}\frac{\partial \delta}{\partial x} \tag{76}$$

This equation can be further simplified by assuming a small axial variation of the liquid film thickness. Using the specific enthalpy h, one can obtain

$$h_g\rho_g v_g + q_{y,g} = h_l\rho_l v_l + q_{y,l} \tag{77}$$

Knowing in these conditions that $\rho_g v_g = \rho_l v_l$ (Eq. (52)) and assuming the interface is impermeable to the species B, one can write

$$\rho_g v_g = \rho_A v_A + \rho_B v_B = \rho_A v_A \tag{78}$$
$$\rho_g v_g h_g = \rho_A v_A h_A + \rho_B v_B h_B = \rho_A v_A h_A \tag{79}$$

or

$$h_g = h_A \tag{80}$$

Eq. (75) becomes

$$q_{y,g} + \dot{m}h_{fg} = q_{y,l} \tag{81}$$

$\dot{m}$ is the liquid evaporation rate and h_{fg} is the latent heat enthalpy.
By substituting the expressions of $q_{y,g}$ and $q_{y,l}$, one can have

$$\dot{m}h_{fg} - k_g \frac{\partial T_g}{\partial y} + \left[\alpha_d R T_g \frac{M^2}{M_A M_B} + (h_A - h_B) \right] J_{Ay,g} = -k_l \frac{\partial T_l}{\partial y} \tag{82}$$

When the Dufour and the interdiffusion of spices effects in the heat flux expression are neglected:

$$\dot{m}h_{fg} - k_g \frac{\partial T_g}{\partial y} = -k_l \frac{\partial T_l}{\partial y} \tag{83}$$

This equation is a simplified condition expressing the energy balance at the liquid-gas interface.

It is important to mention that in the majority of the theoretical works published on the coupled heat and mass transfers in the presence of falling films, numerous assumptions and approximations are retained and used which leads to simple and flexible balance equations on the interfaces.

4. Conclusion

Falling film evaporation is widely encountered in various natural and industrial applications. It encompasses multiple physical phenomena associated with surface tension, shear stress, heat and mass transfer, and others. This book chapter reviews the main studies on falling film evaporation, especially those related to numerical treatment and modeling.

Besides, a frame for the modeling of the fluid flow with heat and mass transfer in presence of evaporation has been established and explained. Therefore, we have presented various aspects related to the formulation of the coupled heat and mass transfer problem with or without falling film. A general interface balance equation was derived and subsequently used to establish the conditions expressing the conservation of energy, mass, and momentum at the interface between a falling liquid and a gas mixture inside confined domain.

Author details

Jamel Orfi[1,2]* and Amine BelHadj Mohamed[3]

1 Mechanical Engineering Department, King Saud University, Riyadh, Saudi Arabia

2 KA CARE Energy Research and Innovation Center, Riyadh, Saudi Arabia

3 Thermal and Energy Systems Studies Laboratory, National School of Engineers of Monastir, Monastir, Tunisia

*Address all correspondence to: orfij@ksu.edu.sa

IntechOpen

References

[1] Jamil MA, Zubair SM. Effect of feed flow arrangement and number of evaporators on the performance of multi-effect mechanical vapor compression desalination systems. Desalination. 2018;**429**:76-87. DOI: 10.1016/j.desal.2017.12.007

[2] Wunder F, Enders S, Semiat R. Numerical simulation of heat transfer in a horizontal falling film evaporator of multiple-effect distillation. Desalination. 2018;**401**:206-229. DOI: 10.1016/j.desal.2016.09.020

[3] Ribatski G, Jacobi AM. Falling-film evaporation on horizontal tubes – a critical review. International Journal of Refrigeration. 2005;**28**:635-653

[4] Qiu Q, Zhu X, Mu L, Shen S. An investigation on the falling film thickness of sheet flow over a completely wetted horizontal round tube surface. Desalination and Water Treatment. 2016;**57**(35):16277-16287. DOI: 10.1080/19443994.2015.1079803

[5] Stephan K. Heat Transfer in Condensation and Boiling. Berlin Heidelberg Gmb H: Springer-Verlag; 1992

[6] Bu X, Weibin MA, Huashan LI. Heat and mass transfer of ammonia-water in falling film evaporator. Frontier in Energy. 2011;**5**(4):358-366

[7] Papaefthimiou VD, Koronaki IP, Karampinos DC, Rogdakis ED. A novel approach for modelling LiBr–H2O falling film absorption on cooled horizontal bundle of tubes. International Journal of Refrigeration. 2012;**35**(4): 1115-1122. DOI: 10.1016/j. ijrefrig.2012.01.015

[8] Raju A, Mani A. Effect of flame spray coating on falling film evaporation for multi effect distillation system. Desalination and Water Treatment. 2013;**51**(4–6):822-829

[9] Faghri A, Zhang Y. Transport Phenomena in Multiphase Systems. Amesterdam, NL: Elsevier; 2006

[10] Rogers JT. Laminar falling film flow and heat transfer characteristics on horizontal tubes. Canadian Journal of Chemical Engineering. 1981;**59**:213-222. DOI: 10.1002/cjce.5450590212

[11] Rogers JT, Goindi SS. Experimental laminar falling film heat transfer coefficients on a large diameter horizontal tube. Canadian Journal of Chemical Engineering. 1989;**67**:560-568. DOI: 10.1002/cjce.5450670406

[12] Zhang JT, Wang BX, Peng XF. Falling liquid film thickness measurement by an optical-electronic method. The Review of Scientific Instruments. 2000;**71**:1883-1886. DOI: 10.1063/1.1150557

[13] Gstoehl D, Roques JF, Crisinel P, Thome JR. Measurement of falling film thickness around a horizontal tube using a laser measurement techniquere. Heat Transfer Engineering. 2004;**25**:28-34. DOI: 10.1080/01457630490519899

[14] Wang XF, He MG, Fan HL, Zhang Y. Measurement of falling film thickness around a horizontal tube using laser-induced fluorescence technique. In: The 6th International Symposium on Measurement Techniques for Multiphase Flows. Vol. 147. 2009. pp. 1-8

[15] Hou H, Bi QC, Ma H, Wu G. Distribution characteristics of falling film thickness around a horizontal tube. Desalination. 2012;**285**:393-398. DOI: 10.1016/j.desal.2011.10.020

[16] Bigham S, Kouhi Kamali R, Noori SMA, Abadi R. Two-phase flow numerical simulation and experimental verification of falling film evaporation

on a horizontal tube bundle. Desalination and Water Treatment. 2015;**55**:2009-2022. DOI: 10.1080/ 19443994.2014.937750

[17] Zhao CY, Qi D, Ji WT, Jin PH, Tao WQ. A comprehensive review on computational studies of falling film hydrodynamics and heat transfer on the horizontal tube and tube bundle. Applied Thermal Engineering. 2022; **202**:117869. DOI: 10.1016/j. applthermaleng.2021.117869

[18] Bu X, Ma W, Huang Y. Numerical study of heat and mass transfer of ammonia-water in falling film evaporator. Heat and Mass Transfer. 2012;**48**:725-734. DOI: 10.1007/ s00231-011-0923-4

[19] Boukrani K, Carlier C, Gonzalez A, Suzanne P. Analysis of heat and mass transfer in asymmetric system. International Journal of Thermal Sciences. 2000;**39**:130-139. DOI: 10.1016/S1290-0729(00)001987

[20] Yan WM. Effects of film evaporation on laminar mixed heat and mass transfer in a vertical channel. International Journal of Heat and Mass Transfer. 1992;**35**(12):3419-3429

[21] Ali Cherif A, Daif A. Etude numérique du transfert de chaleur et de masse entre deux plaque planes verticales en présence d'un film de liquide binaire ruisselant sur l'une des plaques chauffées. International Journal of Heat and Mass Transfer. 1999;**42**(13): 2399-2418. DOI: 10.1016/S0017-9310 (98)00339-1

[22] Yan WM, Lin TF. Evaporative cooling of liquid film through interfacial heat and mass transfer in a vertical channel: Numerical study. International Journal of Heat and Mass Transfer. 1991; **34**:1124-1191

[23] Feddaoui M, Belahmidi E, Mir BA. Numerical study of the evaporative

cooling liquid film in laminar mixed convection tube flows. International Journal of Thermal Science. 2001;**40**: 1011-1020

[24] Agunaoun A, Daif A, Barriol R, Daguenet M. Evaporation en convection forcée d'un film mince s'écoulant en régime permanent, laminaire et sans ondes sur une surface plane inclinée. International Journal of Heat and Mass Transfer. 1994;**37**:2947-2956

[25] Agunaoun A, Ilidrissi A, Daif A, Barriol R. Etude de l'évaporation en convection mixte d'un film liquide d'un mélange binaire s'écoulant sur un plan incliné soumis à un flux de chaleur constant. International Journal of Heat and Mass Transfer. 1998;**41**(14): 2197-2210

[26] Mezaache E, Dagunet M. Etude numérique de l'évaporation dans un courant d'air laminaire d'un film d'eau ruisselant sur une plaque inclinée. International Journal of Thermal Science. 2000;**39**:117-129

[27] Feddaoui M, Mir A, Belahmidi E. Numerical simulation of mixed convection heat and mass transfer with liquid film cooling along an insulated vertical channel. Heat and Mass Transfer. 2003;**39**:445-453. DOI: 10.1007/s00231-002-0340-9

[28] Tahir F, Mabrouk A, Koç M. Heat transfer coefficient estimation of falling film for horizontal tube multi-effect desalination evaporator using CFD. International Journal of Thermofluids. 2021;**11**:100101

[29] Abraham R, Mani A. Heat transfer characteristics in horizontal tube bundles for falling film evaporation in multi-effect desalination system. Desalination. 2015;**375**:129-137. DOI: 10.1016/j.desal.2015.06.018

[30] Jin PH, Zhang Z, Mostafa I, Zhao CY, Ji WT, Tao WQ. Heat transfer

correlations of refrigerant falling film evaporation on a single horizontal smooth tube. International Journal of Heat and Mass Transfer. 2019;**133**: 96-106. DOI: 10.1016/j. applthermaleng.2016.02.090

[31] Alami S, Feddaoui M, Nait Alla A, Bammou L, Souhar K. Turbulent liquid film evaporation in a partially heated wall along a vertical tube. Heat Transfer. 2021;**50**(3):2220-2241. DOI: 10.1002/ htj.21975

[32] Belhadj Mohamed A, Tlili I. Evaporation of a saltwater film in a vertical channel and comparison with the case of the freshwater. Journal of Energy Resources Technology. 2020; **142**(11):103-112. DOI: 10.1115/ 1.4047250

[33] Belhadj Mohamed A, Hdidi W, Tlili I. Evaporation of water/alumina nanofluid film by mixed convection inside heated vertical channel. Applied Sciences. 2020;**10**(7):2380

[34] Ma X, Wang Y, Tian W. A novel model of liquid film flow and evaporation for thermal protection to a chamber with high temperature and high shear force. International Journal of Thermal Sciences. 2022;**172**:107300

[35] Yue Y, Yang J, Li X, Song Y, Zhang Y, Zhang Z. Experimental research on falling film flow and heat transfer characteristics outside the vertical tube. Applied Thermal Engineering. 2021;**199**:117592. DOI: 10.1016/j.applthermaleng.2021.117592

[36] Zhang Z, Wang X, Chen Q, Zhang T. Experimental study on enhanced heat transfer tubes in falling film evaporation. Journal of Physics: Conference Series. 2021;**2021**(2029): 012042

[37] Shahzad MW, Burhan M, Ng KC. Design of industrial falling film evaporators. In: Iranzo A, editor. Heat and Mass Transfer: Advances in Science and Technology Applications. London: InTech Open; 2019

[38] Mahajan RL, Wei C. Buyancy, Soret, Dufour variable property effects in Silicon Epitaxy. Journal of Heat Transfer-Transactions of the ASME. 1991;**113**:688

[39] Gebhart B, Jaluria Y, Mahajan R L, Sammakia B. Buoyancy-induced Flows and Transport. Washington DC, USA: 1988

[40] Bird RB, Stewart WE, Lightfoot EN. Transport Phenomena. New York, USA: Wiley; 2007

[41] Weaver JA, Viskanta R. Natural convection in binary gases driven by combined horizontal thermal and vertical solute gradients. Experimental Thermal and Fluid Science. 1992;**59**(1): 57-68

[42] Hsieh DY, Ho SP. Wave and Stability in Fluids. Singapore: World Scientific Pub Co Inc; 1994

[43] Bel Hadj Mohamed A. Etude du ruissellement d'un film en présence de changement de phase, Msc thesis Report, National School of Engineering of Monastir, 2001

[44] Orfi J. Heat and mass transfer with phase change between a falling liquid film and a flowing gas, 'Transferts de chaleur et de masse avec changement de phase entre un film liquide tombant et un gaz en écoulement' (in French), Post-doctoral degree report, Rapport d'Habilitation de Recherche HDR, Faculty of Sciences of Tunis, University of Tunis-AlManar. 2006

www.ingramcontent.com/pod-product-compliance
Lightning Source LLC
Chambersburg PA
CBHW081229190326
41458CB00016B/5726